MW00712171

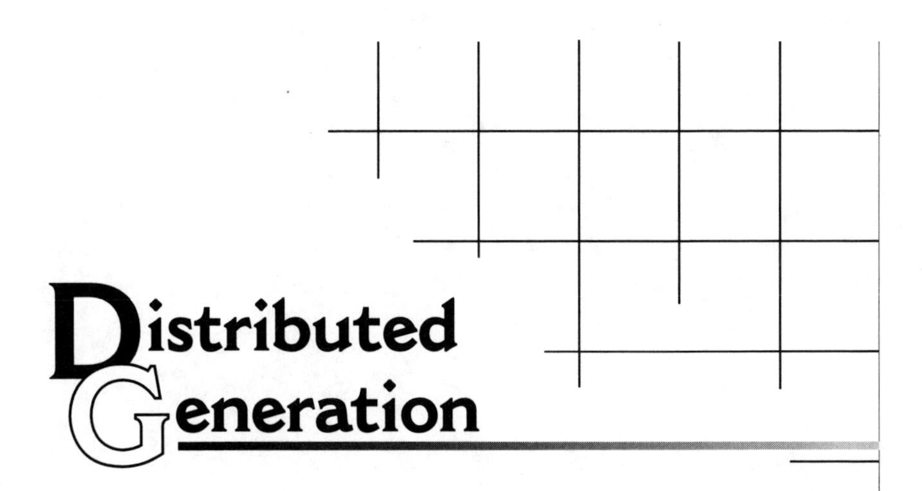
Distributed
Generation

Distributed Generation: A Nontechnical Guide

by Ann Chambers
with Barry Schnoor
and Stephanie Hamilton

PennWell Corporation
1421 South Sheridan
Tulsa, Oklahoma 74112
800-752-9764
sales@pennwell.com
www.pennwell-store.com
www.pennwell.com

Cover and book design by Kay Wayne

Library of Congress Cataloging-in-Publication Data

Chambers, Ann.
Distributed Generation: a nontechnical guide / by Ann Chambers
p. cm.
ISBN 0-87814-789-6
1. Distributed generation of electric power. I. Title.

TK1006 .C43 2001
621.31--dc21

2001021234

Printed in the United States of America

1 2 3 4 5 05 04 03 02 01

For Tony and Aaron
My Bubbas

--Ann Chambers

CONTENTS

List of Figures

List of Tables

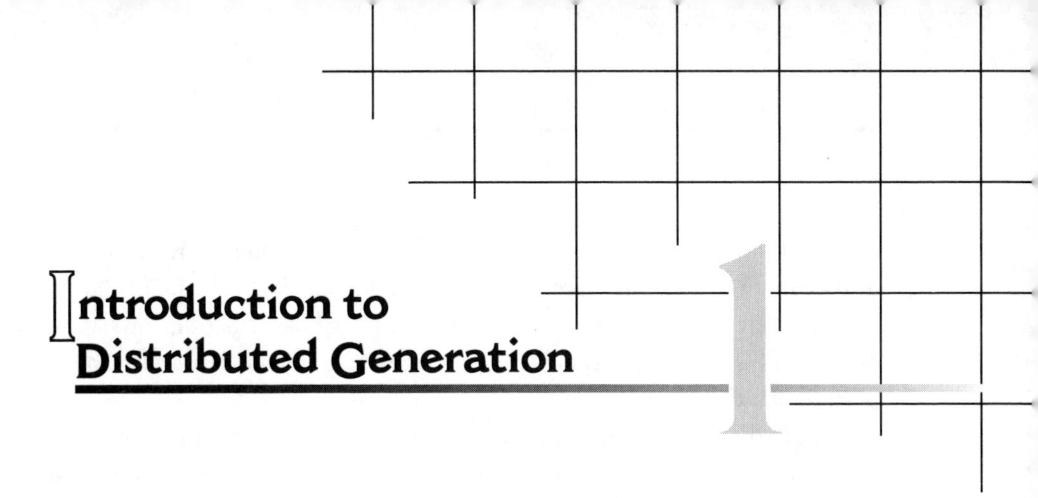

Introduction to Distributed Generation

Today's distributed generation installations are in some ways a return to the early days of electrification. Thomas Edison's first power plants were small installations that illuminated only one or two square miles. Soon, however, Edison's dc power facilities were overshadowed by George Westinghouse's ac facilities that could transmit power over great distances, leading to the utility-scale mammoths that became the mainstay of electric power generation in the United States. The large plants offered great economies of scale and transmitted power over a massive transmission grid. This is the technology that brought affordable electric power to our nation. These facilities ran primarily on fossil fuels. Our nuclear plants are generally even larger versions of this utility-scale plant, with nuclear fuel running the steam generators.

But the changing times have brought changing technologies and economics. Over the past decade or so, the uncertainty of impending deregulation caused utilities to hold off on capital intensive construction projects. This brought narrowing margins of excess capacity as our country's energy use continued to grow. These facts have given birth to the merchant power movement, powered primarily by large-scale gas turbines. But they also have led to the inclusion of smaller technologies in our power generation mix.

Over the past decades, great strides have been made by research and development groups on a great many technologies. Fuel cells first used by NASA received government funding and industry participation for several decades. This technology is now on the verge of commercialization for transportation and stationary power generation.

Similarly, small gas turbines have benefited from the advances in large-scale turbine development, bringing this technology to recent commercial competitive standing. Diesel and gasoline-powered engines, used in transportation, are suitable for a variety of power generation uses and they have certainly made great advances in efficiencies, reliability, and emissions reduction from the transportation industry. These are becoming ever more common in the power generation world. (Fig. 1-1)

Renewable technologies such as wind power, landfill gas, solar, and geothermal are also vying for a portion of today's much needed new power generation capacity. Government assistance in research and, in some cases, tax

Fig. 1-1 *Small modular units need little space and take very little time to install. This JS 100 Euro Silent generation package is equipped with a John Deere 4045 HF 157 Powertech engine. It generates 100 kWe with relatively little noise – 70 dBA at 23 feet.*

Fig. 1-2 *Completion of a NedWind 500 kW wind turbine generator. The plant is expected to generate 2.6 million kWh annually, enough to meet the annual electricity demand of more than 800 households. Wind generation is particularly popular in rural areas, because it can allow farmers to generate additional income from grazing lands while still using the land for farming. The wind turbines use only a tiny fraction of the land they are sited on. Today's wind turbine models are far quieter than previous generations.*

credits or other incentives, help make these technologies more viable.

With the national grid showing its age, and with new transmission lines almost non-existent, distributed generation receives a great boon. These small, generally quiet facilities can be placed next to or near to the customer or customers needing their power. (Fig. 1-2)

Restructuring and Deregulation

Utility restructuring, technology evolution, environmental policies, and an expanding power market are providing the impetus for distributed generation's growth into an important energy option. Utility restructuring opens energy markets, allowing the customer to choose an energy provider, method of delivery, and ancillary services. The market forces favor small, modular power technologies that can be installed quickly in response to market signals. This restructuring comes at a time when the demand for electricity is escalating both domestically and internationally. Impressive gains have been made in the cost and performance of small, modular distributed generation technologies. Regional and global environmental concerns have placed a premium on efficiency and environmental performance. Concerns are growing regarding the reliability and quality of electric power.

A portfolio of small gas-fired power systems is coming onto the market with the potential to revolutionize that market. Their size and clean performance allow them to be sited at or near customer sites for distributed generation applications. These systems often allow fuel flexibility by operating on natural gas, propane, or fuel gas from any hydrocarbon. These include coal, biomass and waste from an assortment of sources including refineries, municipalities, and the forestry and agricultural industries.

Technologies such as gas turbines and reciprocating engines are already making a contribution and they have more to offer through focused development efforts. Fuel cells are entering the market, but need more research and development to see widespread deployment. Also, fuel cell/turbine hybrid systems and upcoming generation fuel cells offer even more potential. (Table 1-1)

Distributed Generation Defined

Distributed generation generally applies to relatively small generating units of 30 MW or less sited at or near customer sites to meet specific customer needs, to support economic operation of the existing distribution grid, or both. Reliability of service and power quality are enhanced by the proximity to the customer, and efficiency is often boosted in on-site applications by using the heat from power generation.

While central power systems remain critical to the nation's energy supply, their flexibility is limited. Large power generation facilities are capital-intensive undertakings that require an immense transmission and distribution grid to move the power.

Distributed generation complements central power by providing a relatively low capital cost response to incremental jumps in power demand. It avoids transmission and distribution capacity upgrades by siting the power where it is most needed and by having the flexibility to send power back into the grid when needed.

	Combustion Turbines	Diesels, Internal Combustion	Fuel Cells	Microturbines	Fuel Cell Hybrids
Applications	On/off grid	On/off grid	On/off grid	On/off grid	On/off grid
Capacity	1-250 MW	50 kW-10 MW	2 kW-2 MW	25-500 kW	250 kW-3 MW
Operating life	40,000 hr	40,000 hr	10,000 hr	40,000 hr	40,000 hr
Capital cost ($)	400-600/kW	500-800/kW	3,000/kW	550/kW	1,500/kW*
O&M cost	5-10 mills/kWh	10-15 mills/kWh	5-15 mills/kWh	5-10 mills/kWh	5-10 mills/kWh
Heat rate (Btu/kWh)	8,000-10,500	9,000-11,000	9,500	12,500	6,000

Source: Edison International
**projected at maturity*

Table 1-1 *Distributed Generation Technology Statistics*

Technological advances through decades of research have yielded major improvements in the economic, operational, and environmental performance of small, modular power generation options.

This emerging group of distributed generation choices is changing the way energy service companies, independent power producers, and customers view energy.

Applications

The main applications for distributed generation so far tend to fall into five main categories:

- Standby power
- Combined heat and power
- Peak shaving
- Grid support
- Stand alone

Standby power is used for customers that cannot tolerate interruption of service for either public health and safety reasons, or where outage costs are unacceptably high. Since most outages occur as a result of storm or accident related T&D system breakdown, on-site standby generators are installed at locations such as hospitals, water pumping stations, and electronic dependent manufacturing facilities.

Combined heat and power applications make use of the heat from the process of generating electricity, increasing the efficiency of the fuel use. Most power generation technologies create a great deal of heat. If the generating facility is located at or near a customer's site, that heat can be used for combined heat and power (CHP) or cogeneration applications. CHP significantly boosts system efficiency when it is applied to mid- to high-thermal use customers such as process industries, large office buildings, and hospitals.

Power costs can fluctuate hour to hour depending on demand and generation availability. These hourly variations are converted into seasonal and daily time-of-use rate categories such as on-peak, off-peak, or shoulder rates. Customer use of distributed generation during relatively high-cost on-peak periods is called peak shaving. Peak shaving benefits the energy supplier as well, when energy costs approach energy prices.

The transmission and distribution grid is an integrated network of generation, high voltage transmission, substations, and lower-voltage local distribution. Placing distributed generation at strategic points on the grid—grid support—can assure the grid's performance and eliminate the need for expensive upgrades.

Stand-alone distributed generation serves the customer but is not connected to the grid, either by choice or by circumstance. Some of these applications are in remote areas where the cost of connecting to the grid is cost prohibitive. Such applications include users that require stringent control of the quality of their electric power, such as computer chip manufacturers.

Customer Benefits

Distributed generation ensures reliability of the energy supply, which is increasingly critical to business and industry. Reliability is essential to some industries where interruption of service creates extremely expensive problems by suddenly shutting down machinery or in industries where health and safety is endangered by sudden outages.

Distributed generation is also able to provide the quality power needed in many industrial applications that are dependent on sensitive electronic instrumentation and controls that cannot withstand power dips or surges.

It can also offer efficiency gains for on-site applications by avoiding line losses and by using both the electricity and heat produced in power generation for industrial processes, heating, or air conditioning.

Customers can benefit by saving on their electricity bill by self-generating during high-cost peak power periods and by taking advantage of relatively low-cost interruptible power rates from their utility.

It allows facilities to be sited in inexpensive remote locations without the need to incur the expense of building distribution lines to connect to the main grid.

Distributed generation increasingly offers an assortment of technologies and fuels, allowing the customer to choose an application that best suits his needs. Also, with each new generation in many technologies, the amount of space needed to house the generation systems shrinks, allowing more flexibility in siting.

Supplier Benefits

Distributed generation limits the capital exposure and risk because of the size, siting flexibility, and fast installation of these systems.

It avoids unnecessary capital expenditure by closely matching capacity increases to growth in demand. It also avoids major investments in transmission and distribution system upgrades by siting the generation near the customer. It also offers a relatively low-cost entry into a competitive market.

It opens the markets in remote areas that do not have an established grid and in areas that do not have power due to environmental concerns.

National Benefits

National benefits of distributed generation include the reduction of greenhouse gas emissions through efficiency gains and through potential renewable resource use. Distributed generation responds to the increasing energy demands and pollutant emission concerns while providing low-cost, reliable energy industry needs to maintain competitiveness in the global marketplace.

Recent technological advances have positioned the United States to export distributed generation to a rapidly growing world energy market, much of which has no transmission and distribution grid.

It is establishing a new industry with the potential to create billions of dollars in sales and hundreds of thousands of jobs. It also enhances productivity through improved reliability and quality of delivered power.

The Market

The coming importance of distributed generation can be seen in the estimated size of the market. Domestically, new demand combined with plant retirements is projected to require up to 1.7 trillion kWh of additional electric power by 2020. That is almost twice the growth of the last 20 years. Over the next decade, the domestic distributed generation market is expected to jump to 5 GW to 6 GW annually to keep up with demand.

Worldwide forecasts show electricity consumption increasing from 12 trillion kWh in 1996 to 22 trillion kWh in 2020. Much of this jump is expected to come from developing countries without national power generation grids. The projected distribution generation capacity increase associated with the global market is estimate at 20 GW annually over the coming decade.

The anticipated surge in the distributed generation market can be attributed to several factors. Under utility restructuring, energy suppliers, not the customer, must shoulder the financial risk of the capital investments associated with capacity additions. This favors less capital-intensive projects and shorter construction schedules. Also, while opening the energy market, utility restructuring places pressure on reserve margins, as energy suppliers increase capacity factors on existing plants to meet growing demand rather than install new capacity. This also increases the probability of forced outages. As a result, customer concerns over reliability have escalated, particularly those in the manufacturing industry.

With the increasing use of sensitive electronic components, the need for reliable, high-quality power supplies is ever more important in most industries. The cost of power outages or poor quality power can be disastrous in industries with continuous processing and pinpoint quality specifications. Studies indicate that nationwide, power fluctuations cause annual losses of $12 billion to $26 billion.

As the electric power market opens up, the pressure for improved environmental performance increases. In many regions of the country, there is near-zero tolerance for additional pollutant emissions as the regions strive to gain compliance. Public policy, reflecting concerns over global climate change, is providing incentives for capacity additions that offer high efficiency and use of renewable energy sources. (Fig. 1-3)

Overseas, the utility industry is undergoing change as well, with market forces displacing government controls and public pressure forcing more stringent environmental standards. There is an increasing effort to bring commercial power to an estimated two billion people in rural areas throughout the globe who are currently without access to a power grid.

Distributed generation is becoming an increasingly popular solution for the future power needs of the United States, primarily because of continuing deregulation of electric power. Tying the merchant power trend to distributed generation allows developers to take advantage of opportunities where traditional utility plants are not the best solution. Large utility plants may sometimes be at a disadvantage in a competitive environment. Big

Fig. 1-3 *Solar arrays such as this one in California are well suited to sunny locales. While the installation cost is relatively high, there is no following fuel cost. A great benefit in areas with air quality concerns, renewable generation from solar or wind power, create no objectionable emissions.*

plants can generate a large amount of electricity at a moderate price, but there are often problems with running these plants at low loads.

Transmission infrastructure construction is becoming more and more of an expense and problem for utilities as well. Distributed generation plants can avoid both problems by installing capacity where it is needed. With distributed generation, a small power generation unit can be placed on-site, or very close, to the facility or facilities that need the power. This eliminates costly overbuilding of capacity and expensive transmission line construction.

The mini-merchant for distributed generation is a new concept, referring to a distributed generation facility that seeks to match its generating portfolio to a local or regional electricity demand profile in the most efficient and economic way. These plants are typically cogeneration facilities, with overall thermal efficiencies as high as 88%. When compared directly to the separate production of electricity and thermal energy, these plant can reduce the CO_2 emission by 50% for the same amount of useful energy. They may also reduce the amount of fuel used by up to 50%.

The mini-merchant plant model hinges on overall economics and how cogeneration and distributed generation fit together. For distributed generation merchant facilities to work well, several characteristics must exist—flexible dispatch, load following, duty cycle, cogeneration, power production, and service territory. These plants can be run on internal combustion engines or gas turbines. (Figs. 1-4 and 1-5)

The electricity production capacity must be capable of being dispatched, cycling on and off based on the price of alternative sources of electricity. To facilitate dispatch, the mini-merchant relies on three classes of generators, responding to base load, intermediate load, and peak load demand requirements. Effective dispatch requires that all engines be capable of starting and synchronizing in less than 30 seconds. In most

***Fig. 1-4** The Wartsila 1,200-rpm 18V220 SG engine provides intermediate load power. It is rated at 2.5 MW.*

Fig. 1-5 *Gas Power System's 1.2 MW Innovator genset can use liquid or gaseous fuels.*

cases, this capability will be unnecessary, but it could be required. Rapid load changes must also be accommodated without tripping off the load and maintenance should not be affected by repeated starting and stopping of the units. These abilities make these small plants far more flexible than standard utility-scale units.

For distributed generation applications, load following capabilities are essential. Reciprocating engine efficiency is reasonably flat between 40% and 100% load for individual generators. By having several engines, it is possible to load follow a local area from base to peak with little effect on efficiency. Large-scale utility plants do not enjoy this luxury. They generally have limited load range for top efficiency.

The difference between baseload and peak averages 100%. For instance, electricity load in the summer months is low at night, when many industrial customers are closed and air conditioners are running very little. But during the day when the industrial customers are operating and air conditioners are cycling, the power demand jumps 100% or more.

To minimize the capital cost for a distributed generation plant, it is important to match the generating equipment type to the expected duty. Peaking requirements are met through peaking generating equipment, intermediate generation is used for intermediate needs and baseload equipment provides for baseload needs.

Thermal energy production, called cogeneration, helps optimize efficiency for distributed generation facilities. Thermal energy production must be reliable with or without electricity production for this ability to truly to be an asset. Natural gas engines have a fairly high exhaust temperature of more than 770 degrees Fahrenheit, corresponding to a plant thermal capacity of more than 24 MWth. Heat is recovered from exhaust gases and used for thermal needs in the facility.

The amount of electricity produced at a cogeneration distributed generation plant or mini-merchant is determined by the size of the thermal host. This ensures that the production efficiency is maintained at an optimum level. When there is little thermal need, all of the generation costs are absorbed by the electricity cost, with none going to a thermal power cost. If electricity is needed at a time when thermal needs are low, the decision to produce electricity versus buying it from outside will depend on a comparison of the incremental cost of production and purchase. Normally the cost of purchasing outside electricity is lowest when weather is moderate. Extremes in climate in both summer and winter increase the electrical demands.

In the open market, there are times when low electricity load conditions on the grid force "must run" facilities belonging to utilities to discount their energy to near zero pricing. When this happens, on-site generating facilities need to have the flexibility to purchase that low cost outside power. The goal of distributed generation, however, is to minimize reliance on the transmission grid for peaking and intermediate generation, and to produce baseload generation when it is economically practical.

Using distributed generation resources sited close to loads allows utilities and other energy service providers to

- provide peak shaving in high load growth areas,
- avoid difficulties in permitting or gaining approval for transmission line rights-of-way,
- reduce transmission line costs and associated electrical losses, and
- provide inside-the-fence cogeneration at customers' industrial or commercial sites.

Homeowner Demand

One million homeowners a year are purchasing backup power systems for their homes, according to figures compiled by Briggs & Stratton. In recent years, Y2K fears, weather patterns such as El Niño and La Niña and their ensuing ice storms, tornadoes, blizzards, hurricanes and heat waves are creating nervous customers looking to ensure their reliability.

The summer outages of 1999 prompted the Department of Energy (DOE) to commission a Power Outage Study Team to evaluate electric reliability. The team's interim report was released earlier this year, predicting that sections of the country will continue to experience serious outages until operations, regulations, and technology can catch up with demand.

There are a multitude of issues that can drive homeowners to backup power systems, including loss of heat, flooded basements when sump pumps lose power, freezer and refrigerator contents spoiling, family members on life-sustaining home medical equipment, and telecommuters who need electronic equipment for their employment.

"I think it is a trend. People want to be protected, particularly those people who are working at home, where going without power for 30 to 36 hours would be a real problem," says Walt Steoppelwerth, known as the "Remodeling Guru." "A lot of builders are now offering entire electrical packages to support all the needs in a home."

Using a permanent transfer system makes a portable generator safer and more convenient for homeowners. The most critical circuits are connected to the generator via the transfer system. Then, if the power goes out, those circuits can be turned on at the transfer switch.

Backup power systems, including a transfer switch and either a 5,000 W or 7,500 W generator and emergency power transfer system, can be purchased for $1,000 to $1,500. They are available at many home improvement, hardware or outdoor power equipment retailers.

Combustion Turbines

Two types of combustion turbines are available for 1 MW to 25 MW distributed generation. Heavy-frame models are relatively rugged with massive casings and rotors. Aeroderivative designs, based on aircraft turbofan engines, are much lighter than the heavy-frame models and operate at higher temperature ratios. They also have higher compression rations, so aeroderivative units have better simple-cycle efficiencies and lower exhaust gas temperatures than heavy-frame models.

Combustion turbine designs typically have dual-fuel operation capability, with natural gas as the primary fuel and a high quality distillate, such as No. 2 oil, as a back-up fuel. Because gas turbines have relatively high fuel gas pressure requirements, a natural gas compressor is usually needed unless the plant happens to be sited near a high-pressure cross-country natural gas pipeline. Combustion turbines typically require a minimum natural gas pressure of about 260 psi, while aeroderivative engines require a minimum natural gas pressure as high as 400 psi. A gas compressor can increase total plant cost by 5 to 10 per cent.

Maintenance costs for heavy-frame units can be about one-half that of aeroderivative units. Major maintenance of heavy-frame units may occur on-site, with an outage of about one week for a major overhaul. With aeroderivative units, the gas generator can be replaced with a leased engine, minimizing the power replacement costs associated with the maintenance outage. Aeroderivative engines can be replaced in two or three shifts, and the removed engine can be overhauled off-site.

Microturbines

The market for microturbine products will be a significant niche, totaling $2.4 billion to $8 billion by 2010, and more than 50 percent of that market will be international. That's one of the conclusions reached by microturbine stakeholders, according to a market forecast from GRI.

Microturbines are of growing interest for distributed power generation because they can deliver combined heat and power, onsite generation, and be the prime mover for refrigeration and air compression. (Fig. 1-6)

Chicago-based GRI used the Delphi approach to conduct its "Microturbine Market and Industry Study." The project is intended to give an expert-based perspective of the market by separating hype from

Fig. 1-6 Unicom Distributed Energy and Honeywell/Allied Signal Power Systems have demonstrated the Parallon 75 microturbine at an energy efficient McDonald's in Bensenville, IL.

economic reality. Thirty-seven experts, representing microturbine manufacturers, utilities, venture capital firms, energy service companies, government entities and other stakeholder organizations were surveyed.

The study concludes that, while initial sales of microturbines will occur primarily in North America, more than half of sales will be international by 2010. Many stakeholders feel microturbines can provide eight percent of the estimated one million megawatts (MW) of new power capacity that will be needed by 2010.

Manufacturers, experts, and utilities believe that the growth rate and market acceptance will be greater internationally, in the long run, because

- fewer barriers are likely to be imposed by existing utilities, power providers or regulators;
- fewer interconnection issues will arise because many applications will be for "prime" power without grid interconnection, and
- shorter value chains will exist, which reduces cost premiums.

Reciprocating Engines

Reciprocating engines vary greatly and have different designs depending on the fuel they burn. Natural gas-fired engines are known as spark ignition or SI engines. Diesel oil-fired engines are known as compression ignition or CI engines. Compression ignition engines can also burn natural gas and a small amount of diesel fuel used as an ignition source. These are known as dual fuel engines.

Distributed generation facilities using reciprocating engines often have several units, rating from 1 to 15 MW each. Medium-speed and high-speed engines derived from train, marine, and truck engines are best suited for distributed generation because of their proven reliability, high efficiency, and low installed cost. High speed engines are generally favored for standby applications, whereas medium-speed engines are generally best suited for peaking and baseload duty.

Reciprocating engines have long been used for energy generators in the United States. However, overseas their ruggedness and versatility have made them popular choices for remote power needs.

Reciprocating engines have a higher efficiency than combustion turbines, although efficiency falls as unit size decreases. Aeroderivative turbines have higher efficiency than heavy-frame combustion turbines in this small size range.

Reliability and availability are important cost-related issues for distributed generation facilities. A 1993 survey found that 56 medium-speed engines at 18 different plants had an average availability of more than 91%. Combustion turbine plants demonstrate availabilities exceeding 95%.

Environmental performance of these technologies depends on what emission is being considered. For NO_x and CO, combustion turbine emissions are 50% to 70% lower than those of reciprocating engines. The NO_x and CO emissions can make it difficult to get permits for reciprocating engines in some states. For CO_2 emissions, reciprocating engines have lower emissions than combustion turbines because of their higher simple-cycle efficiency.

Potential

The worldwide market for distributed generation-size combustion turbines and reciprocating engines has grown in recent years. (Fig. 1-7)

Combustion turbines saw 250 orders in the 1 to 5 MW range in 1997, down from 280 orders in 1996. There were 187 orders in the 5 to 7.5 MW range in 1997, up from 135 orders in 1996. There were 240 orders in the 5 to 15 MW range, up from 49 the previous year.

Reciprocating engines in the 1 to 3.5 MW range saw 4,400 orders in 1997, up from only 1,200 in 1990. There were about 2,100 continuous duty engines sold in 1997, up from 1,300 in 1996. About 370 peaking-duty engines were sold in 1997, down from 870 sold in 1996.

Distributed power systems account for less than 2 GW of electric power, but they are expected to provide as much as 50 GW by 2015.

Fig. 1-7 *The Capstone Model 330 can be grid connected or stand-alone and it can run on natural gas, propane, or sour gas. They are used for load management, back-up power or peak shaving.*

Fuel Cells

Fuel cells are poised to make significant contributions to the growing distributed generation trend. After more than 150 years of research and development, the basic science has been developed and necessary materials improvements have been made to make fuel cells a commercial reality. Phosphoric acid fuel cells, the technology with the earliest promise for large-scale generation, the phosphoric acid fuel cell, is now being offered commercially, with more than 100 200-kW units installed worldwide. More advanced designs, such as carbonate fuel cells and solid-oxide fuel cells, are the focus of major electric utility efforts to bring the technology to commercial viability.

Fuel cells can be described as continuously operating batteries or an electrochemical engine. Like batteries, fuel cells produce power without combustion or rotating machinery. Fuel cells make electricity by combining hydrogen ions, drawn from a hydrogen-containing fuel, with oxygen atoms. Batteries provide the fuel and oxidant internally, which is why they must be recharged periodically. Fuel cells, on the other hand, use a supply of these key ingredients from outside the system and produce power continuously, as long as the fuel and oxidant supplies are maintained. (Fig. 1-8)

The fuel cell uses these ingredients to create chemical reactions that produce either hydrogen- or oxygen-bearing ions at one of the cell's two electrodes. These ions then pass through an electrolyte (which conducts electricity), such as phosphoric acid or carbonate, and react with oxygen atoms. The result is an electric current at both electrodes, plus waste heat and water vapor as exhaust products. The current is proportional to the size of the electrodes. The voltage is limited electrochemically to about 1.23 volts per electrode pair, or cells. These cells then can be stacked until the desired power level is reached. Several stacks can be combined into a "module," for site installation. The waste heat from fuel cells is well-suited for cogeneration or process heat applications.

Fuel cells offer great potential in the distributed generation field for several reasons. Because they are installed in small modules, an industrial customer or utility can install only the amount of power really needed, eliminating extra up-front costs for power that will not be needed for several years. If more power is needed later, more fuel cell modules can be added quickly and easily, with low overhead. The cogeneration and process heat functions of fuel cells are of great appeal to industrial customers, which are currently the great majority of distributed generation providers. Fuel cells also offer very short lead times from order placement to installed power generation capacity.

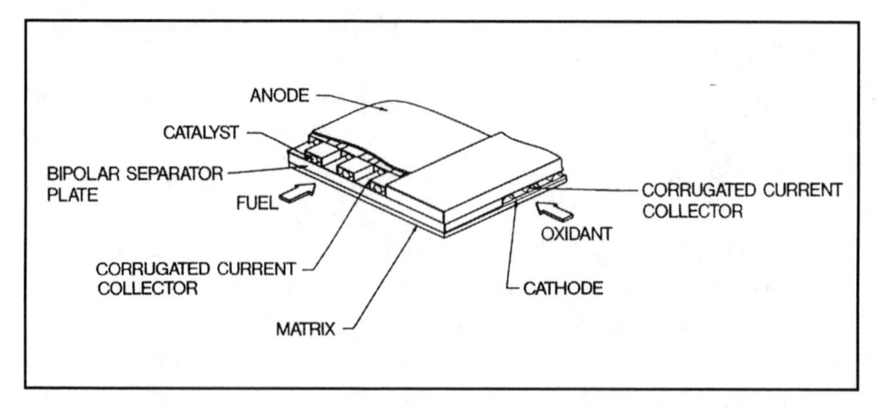

Fig. 1-8 *A Single Cell from a Fuel Cell*

Fuel cell costs have been falling rapidly in recent years, and they should soon become competitive economically with other technologies, especially where strict environmental compliance is required. Operating costs are competitive, particularly when operators consider the fuel cell plant's high efficiency and reliability when operating at partial loads. Siting and operating flexibilities add to their attractiveness, and can translate into site-specific dollar savings.

Internal Combustion

Internal combustion engines can be the prime mover for small power plants, especially for distributed generation or industrial loads. Many of today's installed distributed generation facilities run on these tried-and-true technologies. These engines are basically similar to the gasoline- or diesel-fired engines predominant in today's transportation vehicles. The parts and fuel are readily available, and because there is large commercial demand for these, they are also relatively inexpensive.

History and Current Drivers and Issues

For many years, conventional wisdom and applications subscribed to the "economies of scale" approach for new power plant construction. Bigger plants meant lower unit costs in dollars per kilowatt hour. Fossil-fired units with capacities greater than 1,000 MW were built in the 1970s and 1980s. For high load growth conditions, investment in such large units seems justified. As load growth slows, however, the tide turns toward smaller units. The emergence of the independent power producer and the development of highly efficient generation technologies for smaller loads, has resulted in a growing awareness and interest in smaller-scale, but still competitive, generating plants. One-of-a-kind, large-scale plant design and construction is making room for high-volume, assembly-line production of distributed generation assets.

In the early days of electricity, distributed generation was the rule, not the exception. Without interconnected networks of electricity transmission and distribution lines, end users who wanted the benefits of electricity had to generate it themselves. Millions of farmers, for example, turned to windmills to power their pumps and lights. As distribution networks grew, distributed generation declined, succumbing to the convenience and often, the lower cost, of grid-supplied electricity. Several factors are now converging that may reestablish distributed generation as a significant force in power generation.

First, centralized, command-and-control regulation of the electricity is giving way to a much more market-driven approach. Countries on every continent are privatizing their electricity industries, increasing the viability and attraction of distributed generation.

Second, as generation assets are transferred from government or utility ownership to third-party control, asset management and low cost operation assumes critical importance. Asset owners will reexamine their portfolios to assemble the most effective combination of central station and distributed resources.

Third, the competition that is unleashed through privatization will empower consumers to seriously consider generation alternatives. Distributed generation will benefit both from utilities that invest in distributed generation to retain key customer loads and from customers that invest for self-generation or combined heat and power.

Distributed generation is attractive to utilities because it can avoid the siting and construction of expensive power lines and substations, and because it can allay concerns surrounding electromagnetic fields and the environmental impacts of large transmission towers.

Based on assumptions in ABB's "Introduction to Integrated T&D Planning," it can cost $365 to $1,100 per kilowatt hour to run a six-mile power line to a 3 MW customer. A distributed generation system can potentially satisfy the same load at $400 to $500 per kilowatt hour. It may also reduce the energy lost during long-distance transmission, which can be up to 5% in the United States and as much as 30% in developing countries. (Table 2-1)

The installation of diesel and gas-fired generators along transmission lines is also being used to provide backup power in the event of outages.

Distributed resources can also be used to free up transmission assets for increased wheeling capacity. During periods of high electricity demand, distributed generation resources could supply a portion of the local demand, enabling excess capacity to respond to greater revenue opportunities.

Forecasts for distributed generation are promising. Several U.S. experts predict that 10% to 30% of new generating capacity will be met with distributed generation, representing a potential market of several tons of GW. Overseas opportunities are significant as well. Combined heat and power

Technology	Capital Costs
Advanced Combustion Turbine	465
Conventional Combustion Turbine	332
Advanced Combined Cycle	580
Conventional Combined Cycle	449
Gas/Oil Steam Turbine	1,012
Scrubbed Coal, new	1,102
Integrated Gas Combined Cycle	1,315
Fuel Cells	2,163
Advanced Nuclear	2,390
Biomass	1,877
Solar Thermal	3,059
Solar Photovoltaic	4,836
Wind	1,086

Overnight capital costs include just the costs of permitting and building plant – they do not include any interest expenses

Source: Energy Information Administration

Table 2-1 *Overnight Capital Costs by Technology ($/kW)*

plants are increasing in popularity around the world, and village power electrification programs are widespread throughout Africa, Asia, and South America. More than $30 billion per year is spent on home heating fuels, batteries, and candles in rural parts of the world, along with annual investments in 10 to 15 GW of diesel generating capacity.

Distributed generation aims to capture a large fraction of this rural market. One of its greatest attractions is its short development schedule. The time needed to identify a potential project, develop and negotiate for the site and permits, arrange financing, build the project, and finally test and start it varies widely by type of project. A utility-scale coal-fired power plant can take five to seven years in development. A large natural gas-fired generating facility can take anywhere from three to five years. Distributed generation can run the gauntlet in only six to 18 months.

Advances in technology have lowered the size threshold for power generation facilities to be economically viable, especially when considering microturbines and fuel cells. Power electronics technology is also rapidly evolving, decreasing the cost of protective relaying, interconnection, and remote communications.

The ready availability and competitive pricing of natural gas in recent years has opened the door to a landslide of new gas-fired merchant power plants, and also to the infiltration of smaller gas turbines for distributed generation uses.

Environmental regulations are becoming ever more stringent, adding expenses to many of the older utility facilities and making the new distributed generation technologies more attractive.

Increased sensitivity to power quality is driving many medium and large commercial customers plus many industrial customers to install or at least consider installing onsite generation equipment. These installations can be prompted by reliability needs or by power quality needs. For some manufacturing facilities, if the incoming power even flickers, it can cause equipment to stop, ruining all product that is current in the manufacturing process, and thus costing the company millions of dollars in lost product and lost time. These facilities are ideal candidates for onsite power generation and backup power installations.

Distributed generation can hold great appeal to utilities as well as consumers. Utilities can use distributed generation to create business ventures and alliances, to boost system reliability and customer satisfaction, and to defer capital expenditures. Distributed generation installations can help balance supply and demand, defer construction of transmission and distribution infrastructure, delay need for larger generation facilities, reduce line losses in the transmission and distribution system, and assist in delivery of ancillary services.

Several strategies are available to utilities interested in making distributed generation work *for* rather than *against* them. Utilities can build or install new distributed generation projects at customer sites. They can build or install distributed generation projects on their own transmission or distribution systems. They can use existing customer onsite generation capacity to meet peak loads on the grid. They can also rent or lease distributed generation resources from third-party developers.

The benefits from such activities can be significant. For the utility to benefit, however, it must control the asset.

Standby

A unique variation on distributed generation is the use of existing emergency and standby generators to supplement electricity supplies during peak demand periods. Many power-critical facilities such as hospitals and manufacturing plants maintain standby generators in case of power outages. As such, they are typically used only a few times per year, so they represent underutilized capital.

If these auxiliary generators could be cost-effectively integrated and controlled on the grid, the asset owned would gain a revenue source and the grid would benefit from the availability of additional capacity to satisfy demand peaks.

Interconnecting a series of auxiliary systems into the grid requires sophisticated dispatch and control mechanisms. The standby generators must be synchronized into the grid when they go on-line, necessitating close control over generator status. The dispatch system must analyze the participating generators and grid loads and make quick decisions as to which generator will satisfy the incremental load without upsetting the system.

Finally, the standby generators must possess high availability factors, dictating a rigorous maintenance program to ensure reliability.

Several companies are pursuing this concept with test programs, integrating auxiliary generators into the grid. Encorp developed a software system for the remote monitoring, control, and grid paralleling of dispersed generators. Retrofitting standby generators with closed-transition switchgear and equipment to ready them for remote control frees capacity for peak shaving revenue opportunities. Capital costs for retrofitting existing systems are estimated to be around $100 per kilowatt hour, well below the cost of building new capacity.

Growth Trends

The U.S. industrial sector, which accounts for about one-fourth of the nation's total energy use, and much of its distributed generation load, is expected to increase its demand for energy by 1.3% annually over the next couple of decades, resulting in more than a 25% jump in that sector's demand by 2018.

The Gas Research Institute projects that total industrial energy consumption will grow from 27.3 quadrillion Btu (quads) in 1995 to 35.1 quads in 2015. During that same period, industrial consumption of natural gas will increase from 10 quads in 1995 to 13 quads in 2015. Natural gas has a dominant share of industry's competitive fuel and power segment, at 40%, and this share is expected to be maintained during the projection period. The market includes boilers, industrial cogeneration, and process heat.

In the competitive fuel and power markets, natural gas is expected to increase its share of the boiler market, at the expense of coal; grow in the cogeneration markets as a reflection of end user preference for natural gas, combined-cycle technologies, and maintain its dominant share of the process heat sector.

The Challenge

Although growing, distributed generation is still in its infancy. Ultimately, the market will be shaped by product development and economic, institutional, and regulatory issues.

Market penetration will depend on how well manufacturers of distributed generation systems do in meeting product pricing and performance targets. Many of the more promising technologies have not yet achieved market entry pricing or risk levels, while others simply have not reached their market potential.

Customers have yet to define and quantify distributed generation attributes such as transmission and distribution upgrade cost avoidance, improved grid stability, or enhanced power reliability.

A major institutional issue, regarding customer interconnection with the distribution grid, currently stands in the way. Utility specifications for connection with the grid are complex and inconsistent. The results are high costs and project delays, or termination. Interconnect requirements are needed for safety, reliability, and power quality. This suggests that national interconnect standards are needed. Other costs that need addressed include, historical use charges, back-up charges, insurance charges, and other utility fees associated with choosing to self-generate while remaining connected to the grid.

Regulatory issues arise as well. For example, unless changes are made, distributed generation units may not receive credit for avoided pollutant emissions. These emission credits are normally dealt with during the utility resource planning process.

For distributed generation to reach its potential, the technical, economic, institutional, and regulatory issues must be dealt with effectively. This will require cooperation between the public and private sectors. If this happens, a new industry can blossom, bring jobs and revenues.

Reliability

The DOE has issued 12 recommendations in response to the troubling outages during peak demand in 1999. The recommendations come from the Power Outage Study Team (POST), which included experts from DOE, national laboratories, and the academic community. The team examined several of last summer's events in detail and recommended action to help prevent similar problems in the future. The DOE contends that this study is differentiated by its consideration of industry restructuring.

"The power outages and disturbances studied by POST served as a wake-up call, reminding us that reliable electric service is critical for our health, comfort, and the economy," the DOE report states. "POST conducted a thorough study of eight outages and disturbances in different parts of the country. The team visited the sites and interviewed the people who were operating the systems at the time of the occurrences, obtaining valuable first-hand information and data."

Many of the recommendations included in the report are intended to address reforms needed to allow the restructured markets to fulfill their potential for improved reliability. Other recommendations address the importance of continuing to invest in the development and use of tools needed for monitoring and maintaining the electric system and for responding to system emergencies.

The final report, issued in March 2000, lists a dozen recommendations for consideration by the Secretary of Energy, along with possible federal actions. Recommendations include the following:

Promote market-based approaches to reliable electric services. The value of reliability needs to be determined in competitive markets, and customers, as well as energy providers, need to have the opportunity to participate in markets for energy and ancillary services. Federal action recommended: support the implementation of fair, efficient, and transparent markets for electric power and ancillary services.

Enable customer participation in competitive electricity markets. To more fully participate in a competitive market, customers need to see real-time prices if they wish to have access to the communication and control technologies that will enable them to participate directly. Federal actions recommended: support development of market rules allowing customers to supply load reductions and ancillary services in competitive energy markets, and encourage development of demand management systems that support electric reliability.

Remove barriers to distributed energy resources. There is great interest in distributed generation technologies as a way for utilities to respond quickly to an increased demand for electricity where demand is already high. At the same time, utilities are working to improve power quality. Federal actions recommended: support development of interconnection standards for distributed energy resources, support state-led efforts to address regulatory disincentives for integrating customer supply and demand solutions, and study the potential for using emergency backup generators to reduce system demands to help avoid outages.

Support mandatory reliability standards for bulk-power systems. The grid is being transformed from one that was designed to serve customers of full-service utilities, each integrated across generation, transmission and distribution functions, to one that will support a competitive market. Federal action recommended: support the creation of a self-regulated reliability organization with federal oversight to develop and enforce reliability standards for bulk-power systems as part of a comprehensive plan for restructuring the industry.

Support reporting and sharing of information on best practices. Although there are many forums available for exchange of best practices information, there is concern regarding the consistency and availability of data. Federal actions recommended: promote use of uniform definitions and measurements for reliability-related information, facilitate collection and sharing of information on reliability-related regulatory issues among state public utility commissions, and support activities to develop and share information among industry participants on critical resources and industry practices.

Enhance emergency preparedness activities for low-probability, high-consequence events on bulk-power systems. Last summer's events showed that effective communication and coordination among many parties are critical during system emergencies. Federal action recommended: work with regional, state, and local authorities to support continuous improvement in coordination, planning, and preparations to respond to electricity emergencies.

Demonstrate federal leadership through promotion of best reliability practices at federal utilities. Federal utilities have long pursued many federal and regional objectives, and as part of their role, these utilities have served as research and development catalysts for technological changes in the industry. Federal action recommended: develop and pilot reliability self-assessment procedures, support distributed energy resources and encourage economic energy efficiency.

Conduct public interest reliability-related R&D consistent with the needs of a restructuring electric industry. Industry investments in reliability-related R&D have declined steadily over the past few years. POST supports Secretary Richardson's commitment to increase federal investments in electric reliability R&D in the 2001 budget. Federal actions recommended: develop real-time system monitoring, communication and control technologies; create sensors, remote monitoring and diagnostic technologies for cables and aging transmission and distribution infrastructure; integrate customer demand management, distributed generation and storage technologies; improve analytic models for load forecasts, power system simulations and contingency assessments; and examine the design and performance of competitive electricity markets.

Facilitate regional solutions to the siting of generation and transmission facilities. Stable incentives for investing in generation and transmission must be complemented by siting boards that can discharge their responsibilities in a timely and coordinated fashion. Federal actions recommended:

convene regional summits to initiate and facilitate dialog among regional stakeholders, support federal legislation to facilitate state efforts to form regional siting boards, and raise reliability as an issue as appropriate whenever federal permits are required for siting electric facilities.

Promote public awareness of electric reliability issues. General public awareness of the complex issues associated with maintaining reliable electric service is low, but public interest in reliable service is high. Greater understanding will lead to better-informed decisions. Federal actions recommended: continue DOE-sponsored independent investigations of significant power outages and reliability events, and continue DOE-sponsored forums where stakeholders can meet to discuss reliability issues.

Monitor and assess vulnerabilities to electric power system reliability. The outages studied resulted from unpredicted events that exploited specific weaknesses in physical systems and the planning and operating processes that supported them. Known and emerging electric system vulnerabilities need to be studied from a national perspective by the federal government in close partnership with the utility industry. Federal action recommended: work with industry members to conduct comprehensive assessments of system vulnerabilities and work with industry members to refine and implement procedures to assess the robustness of the system in responding to bulk-power system emergencies.

Encourage energy efficiency as a means for enhancing reliability. Energy efficiency measures can boost system reliability by reducing demand growth in areas experiencing shortages in generation or constraints in transmission or distribution. Technologies and practices that reduce loads during peak demand, such as high-efficiency air conditioning and lighting equipment, are especially valuable. Federal actions recommended: work with state and local governments to support development and implementation of cost-effective energy efficiency programs and expand existing federal programs to promote energy efficiency.

Fuel Cell Market Escalates

The fuel cell market in 1998 was approximately $355 million, but with a projected average annual growth rate (AAGR) of 29.5%, the market is forecast to reach $1.3 billion by 2003, according to the "Fuel Cell Industry Review" from Business Communications Company Inc.

While the fuel cell market has been growing steadily over the past several years, two major obstacles to their widespread use persist—high manufacturing costs and relatively short operating lives. These problem areas are the focus of current research and development efforts.

The fastest growing type of fuel cell is the phosphoric acid category, expected to leap from $41 million in 1998 sales, to $250 million in 2003. (Table 2-2) The largest market is for solid polymer fuel cells, which has an estimated $80 million in 1998 sales. Solid polymer sales are expected to top $450 million in 2003.

	1998	**2003**
Solid Polymer	80	450
Molten Carbonate	100	350
Phosphoric Acid	41	250
Solid Oxide	117	185
Alkaline	17	60
Total	355	1,295

Table 2-2 *Fuel Cell Sales by Type ($ Millions)*

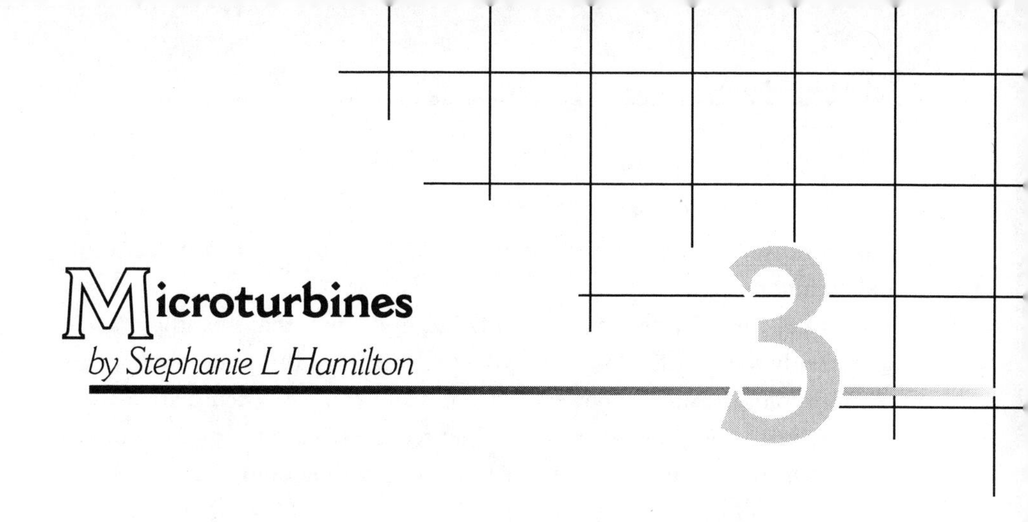

Microturbines

by Stephanie L Hamilton

Microturbine Generators (MTGs) are emerging as new products for small stationary power. MTGs are beginning to prove economically viable and reliable sources of electricity for selected businesses and facilities. Since 1995, a combination of events – technology advances, regulatory changes, and economic and environmental concerns – have allowed these small power plants to finally come out of R&D, into beta-prototypes, and now commercial production. Although they still face some hurdles, MTGs are being seriously considered for use in commercial and other applications.

In this chapter we explore this exciting new technology. We will examine its opportunities for independently powering businesses, or working in parallel with the existing grid as an integrated component to serve commercial and other customers, both in the United States and overseas.

MTGs match the size and load profile of commercial and small industrial customers – who tend to have fairly constant loads over their periods of operations. MTGs perform best when running at high load factors and under constant load. Commercial and small industrial customers, in general, tend to provide this kind of load factor and a constancy of load.

MTGs range in size from 25 kW to 300 kW, although applications larger than 100 kW are not expected to be commercially available for several years because they are in the process of being scaled up for manufacturing and commercialization.

One of the exciting aspects of MTG technology is that it is still in a relatively early stage of development and deployment. At this point other ancillary technologies, such as power electronics and control systems, are maturing. Improved components, such as better recuperators, and advanced materials, such as ceramics, are expected to bring about even more advances through MTG development programs funded by the U. S. Department of Energy (DOE) and others.

MTGs are becoming recognized as technologically sound and economically feasible in certain applications because of several expected desirable characteristics:

- Competitive economics in some applications and markets
- Ease of installation, low maintenance, and simple operability
- Modularity and use of multiple units for potentially enhanced reliability
- Low NO_x emissions
- Compact size, with a footprint of 15 square feet for some designs

Table 3-1 shows an economic comparison of a variety of distributed generation technologies. It is clear that MTGs require additional improvement to compete with currently entrenched diesel and gas engine generators and future new technologies, such as fuel cells. Yet, part of the excitement is that small local power plants, in a variety of technologies, are beginning to offer an on-site alternative to large and remote central power plants, particularly in growth areas.

MTGs are expected to remain easy to install. When natural gas is the fuel for the MTG and the customer desires to be grid connected, then the MTG will require installation on the gas pipeline and the grid. Installation includes both require physical connections and contractual arrangements must ensure that operational and safety requirements are met and that the financial transactions can be captured and consummated. Additionally, contractual arrangements must ensure that all parties know the expected operations of the parties and their rights and obligations.

Economics Comparison of Distributed Generation Technologies

Technology Comparison	Diesel Engine	Gas Engine	Simple cycle Gas Turbine	Microturbine	Fuel cell	Photovoltaics
Product Rollout	Commercial	Commercial	Commercial	1999-2000	1996-2010	Commercial
Size Range (kW)	20-10,000+	50-5,000+	1,000+	30-200	50-1,000+	1+
Efficiency (HHV)	36-43%	28-42%	21-40%	25-30%	35-54%	n.a.
Genset Package Cost ($/kW)	125-300	250-600	300-600	350-750*	1,500-3,000	n.a.
Turnkey Cost-No Heat Recovery ($/kW)	350-500	600-1,000	650-900	600-1,100	1,900-3,500	5,000-10,000
Heat Recovery Added Cost ($/kW)	n.a.	$75-150	$100-200	$75-350	Incf.	n.a.
O&M Cost ($/kW)	0.005-0.010	0.007-0.015	0.003-0.008	0.005-0.010	0.005-0.010	0.001-0.004

*Commercial target price
Source: Distributed Generation Forum managed by Gas Research Institute (GRI), 1999

***Table 3-1** Economic comparison of a variety of distributed generation technologies.*

Most models of MTGs do not require a transformer. A transformer adds to the installation requirements and increases the cost. Most new MTG models offer an integral gas compressor, which allows for the MTG to be installed in a larger variety and number of locations than would be possible if there were no option to purchase the compressor.

Likewise, most MTGs either offer a combined heat and power (CHP) option or have CHP integrally built in.

Some MTG models have demonstrated very easy maintenance – in some cases, limited to changing of air filters after 8,000 hours of operation. The next maintenance will be at 16,000 hours of operation, which will include changing of the air filters and replacement of some consumables, such as igniters.

Most MTG models offer a convenient operator's panel for easy operations. An operator's manual and error codes help with any troubleshooting and ultimate problem resolution. Experience with MTGs shows that early production models have experienced more problems than more mature products. There can be extreme variation in overall maintainability and operability depending on the maturity of the product; in fact, some early production products may prove to be not operable at all.

MTGs that come in small kW output sizes can be bundled together to provide generation reliability, such as matching a customer's power load requirement of five MTGs to the bundled MTG output of seven MTGs. This arrangement can result in reliability that can be quite high if the MTGs can perform reliably. Some

MTGs have demonstrated such highly reliable performance.

Although not all manufacturers are claiming NO_x less than 9 ppm, corrected to 15% O_2, some have already demonstrated such capability. This less-than-9-ppm level, as required by the South Coast Air Quality Management District (SCAQMD), the authority in the Los Angeles basin that regulates air emissions, is expected to set the trend, if not the requirement, in the MTG industry.

Fig. 3-1 *A Capstone Micro Turbine*

When you first see an MTG, you may be surprised at how small it is. Some MTGs are not much larger than a freezer or large refrigerator. Figure 3-1 shows a Capstone Microturbine. These small, compact packages weigh from 1,000 to 3,000 pounds, depending on the customer-selected options and require about 15 square feet of space for the MTG and its necessary clearance area.

The design of the MTG is a major factor contributing to the commercial economics of MTGs. Use of conventional materials, such as steel in the turbine blade, high volume manufacturing advantages, advanced power electronics, and the use non-liquid lubricating methods in some designs, are factors in making the costs of the MTG competitive.

The MTG turbine, operating at high speed, allows relatively high power generation from a small turbine. This enables manufacturers to efficiently use materials and reduce costs. This contributes to their ability to produce MTGs that may rapidly become economically competitive with other forms of small stationary power.

In general MTGs have the following design characteristics:

- For all but one design, only one major moving part
- Most designs boast 40,000 hour "wheel" life; one claims 80,000 hours
- Single-stage compression with a recuperator to boost efficiency
- Advanced power electronics
- Use of conventional steel in turbine wheel
- Some designs make use of innovative air bearings

Most MTG designs have only one major moving part that stacks several components compactly on one rotating shaft. Other small power plants, such as diesel generator sets and other reciprocating engines, have many moving parts. Although current operating efficiencies and overall economics of these alternative generators are formidable, many moving parts results in a more complex design, and creates greater opportunity for any to fail and shut down the machine. Figure 3-2 shows a cutaway of a Capstone MTG turbine and assembly package. It is generally typical of the industry for a single shaft MTG.

The wheel life of the MTG is a major deciding factor in determining how long the MTG will last before a major overhaul is required. An expected life of 40,000 to 50,000 hours for the turbine will make it relatively competitive with diesel generators, now in commercial use. Most manufacturers plan to replace the turbine near the end of the expected five-year life by simply pulling the turbine and replacing it. "Pulled" turbines will be refurbished and used as spares. The expected life is yet commercially unproven. Success of this industry relies on this life expectation being met. It will be interesting to see how well this expectation is met by the industry.

MTGs are single-stage turbines. Single stage turbines are lower in efficiency than multi-stage turbines. To increase the turbine efficiency, a recuperator is added to recover exhaust heat and, in steady state operation, used to raise the temperature of the inlet air to the MTG. Increasing the air inlet temperature reduces the amount of fuel needed and thereby increases the overall efficiency of the MTG. It is estimated the efficiency is increased from about 19% to about 27% as a result of adding recuperation.

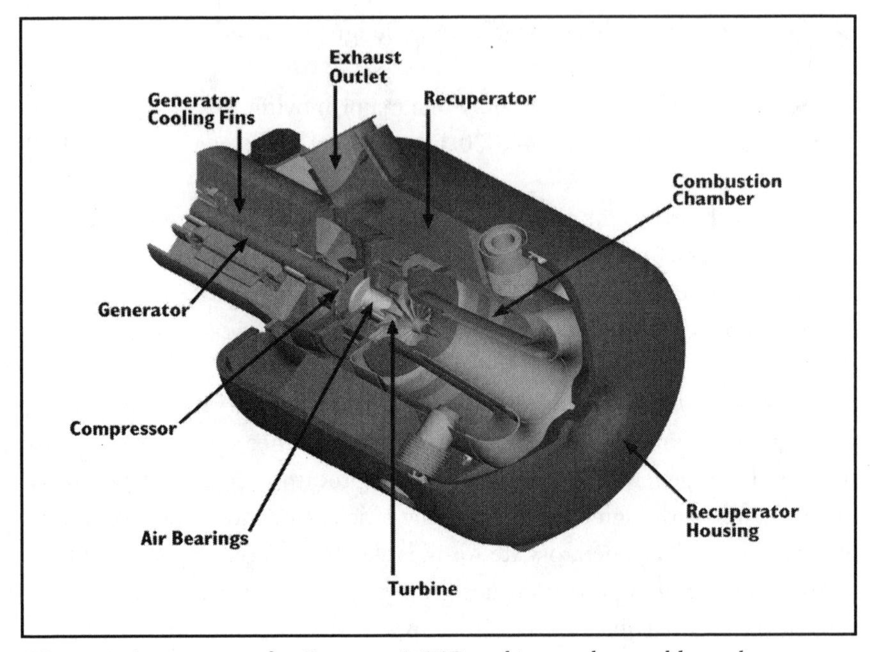

***Fig. 3-2** A cut away of a Capstone MTG turbine and assembly package.*

Manufacturers keep the MTG cost down by using conventional steel turbine wheels. Presently, the use of ceramics and more exotic metals would increase the costs of MTGs. However, the use of conventional steel limits the MTG's operating temperatures, which generally limits the maximum efficiency achievable. To date, the economic trade-off has been to use conventional materials and provide lower efficiency.

The ability to operate the MTG in various ways gives the customer more options in satisfying his/her energy needs and running his/her operations as desired. In general MTGs have the following expected desirable operational options:

- Part-load performance capability
- Alternative fuel and dual fuel operations
- Stand-alone, grid-connected only, and grid parallel with "black start"
- Combined heat and power

The ability to handle a partial load has been demonstrated by some MTGs. However, in the particular demonstration there was a noticeable loss in MTG efficiency under part-load operations.

MTGs have demonstrated alternative fuel capability. It is expected that MTGs will be able to use dual fuels. For example, Capstone has demonstrated the capability of running on either pipeline quality natural gas, which is about 1,050 MMBtu/cf, "ultra rich landfill" type fuel, having an energy content of about 700 MMBtu/cf, as well as using the same combustor.

Elliott Energy Systems' specification shows the capability to use diesel or natural gas at high pressure.

Depending on the relative energy content differences in each fuel, operations on some fuels may require a combustor changeout prior to a change in the type of fuel to be consumed by the MTG.

Most recently available MTGs offer the valuable capability of providing stand-alone operation. Stand-alone operation requires no grid interconnection although having grid interconnection is desired by most customers. Stand-alone operational capability allows a grid-interconnected customer to have MTGs as back up power or emergency power and operate with no power from the grid, if the customer chooses to do so. The customer will need to be able to sustain the required time to "ramp up" the MTG to full power. If this several-minute interval is acceptable, then MTGs can provide this capability.

Also, the customer can use stand-alone operation to provide power during periods when power prices spike and are passed through directly to the customer. The customer could sell power back to the grid and receive the differential prices, as well.

Finally, some customers will have specific thermal and power requirements. MTGs can meet thermal and power requirements if such equipment is selected by the customer. To the extent that the thermal and power requirements can be matched with the capability of the MTG, this capability can be quite advantageous economically for the customer.

Although new large combined-cycle, gas-fired central plants can achieve efficiency as high as 55% to 60% and offer very good emission profiles, these plants also require extensive effort and long lead times to perform necessary funding, planning, engineering, installation, permitting, and start-up. Compared to large central plants, MTGs have several significant advantages.

These advantages include the following:

- Ease of installation
- Simpler siting and licensing
- Lower capital requirements
- No need for large transmission lines
- The reduction of system energy losses

Until 2000 and the experience in California and throughout the West, most experts believed that MTG sales and installations would increase in the United States and North America but would grow faster outside the United States except in California and the West generally. In the United States, bulk power prices, particularly off-peak, are very low in most markets, and the overall reliability of the U.S. grid is high. Given these conditions, customers large enough to use MTGs will want to have reasons beyond providing "bulk power" only to take on the added requirements of being in the "power" business. In fact, most potential MTG users will want the financial benefit of installing and using on-site generation to more than offset the cost and responsibility of operating their own generating plant.

Some experts believe small commercial and light industrial customers will be receptive to energy service providers, who will provide service like a utility. Such services must be at comparable or better reliability than the utility can offer, given regulatory constraints. Service providers should be prepared to, and capable of, handling the capital funding required, if desired by the customer.

Changes in electric industry structure have created opportunities for MTGs. Additionally, worldwide deregulation and privatization give customers choice among energy providers just as emerging technologies give customers the ability to choose their energy source. As MTGs prove out their manufacturers' claims, and manufacturers also further advance their products, MTGs will fill market niches that prize its attributes. Successfully filling one market niche will be a basis for future niche expansion.

Some in the industry believe that MTGs and other forms of distributed generation, when dispersed along the grid, will offer improved reliability to the grid. This belief has not yet been proven nor has the condition been physically tested nor demonstrated. Much work needs to be done.

First, work needs to be carried out to assure that the operation of small generators distributed as integral components on the grid does not degrade the existing high reliability of the grid. Secondly, even more work must be done to ensure that distributed generation improves grid reliably rather than making the grid unstable and causing disruptions and power outages.

There are many companies entering the MTG marketplace. The following six companies are the leaders and the best known. They are listed in alphabetical order.

Bowman Power Systems. Bowman Power Systems (BPS) is located in Southampton, England. It was founded in 1994. BPS adds balance of plant to its TURBOGEN family of small-scale compact power generation systems ranging in size from 25 kW to 250 kW for distributed power generation and for mobile power applications. BPS specializes in CHP and its MTG integrally provides CHP to provide customers with improved economies as compared to power-only applications. CHP applications can have efficiencies in the 75-85% range as compared with 20%-30% efficiencies of power-only applications. BPS is now producing second and third generation MTGs.

According to BPS, it is a leader in design and supply of turbo alternators, power conditioners, and gas boost compressors to the MTG industry.

Figure 3-3 shows a BPS 80 kW CHP MTG. Table 3-2 are the features of the BPS 80 kW. Figure 3-4 is a drawing of the BPS System Arrangement.

Fig. 3-3 *A BPS 80 kW CHP MTG.*

Standard Features

- Turbo alternator - 2 bearings integrated mono-rotor design
- Engine - micro gas turbine rated for continuous operation
- Low emission combustor
- High effectivity recuperator
- Alternator - High speed, 4 pole permanent magnet
- Base frame
- Acoustic canopy - 75 dBA @ 1 meter
- Electric starting AC or DC
- Power conditioner
- Package mounted control
- LCD display
- Output circuit breaker

Available Options

- Gas safety fuel train
- Gas boost compressor
- Ultra low noise enclosure - 65dBA @ 1 meter
- Parallel operation with other generators
- Utility feed interface
- Battery or utility starting
- Remote control data logging (local and/or remote)

Contact Information

Bowman Power Systems Inc.
Tony Hynes
20501 Ventura Boulevard, Suite 285
Woodland Hills, California 91364 USA
phone: 818.884.1944
fax: 818.884.0991
email: la.bowmanpower@att.net

Bowman Power Systems Ltd.
Dave Streather
Ocean Quay, Belvidere Road
Southampton, Hampshire SO14 5QY England
phone: (within UK) (0)2380-236-700
(outside UK) 44-2380-236-700
email: dstreather@bowmanpower.co.uk

www.bowmanpower.com

***Table 3-2** Features of the BPS 80 kW.*

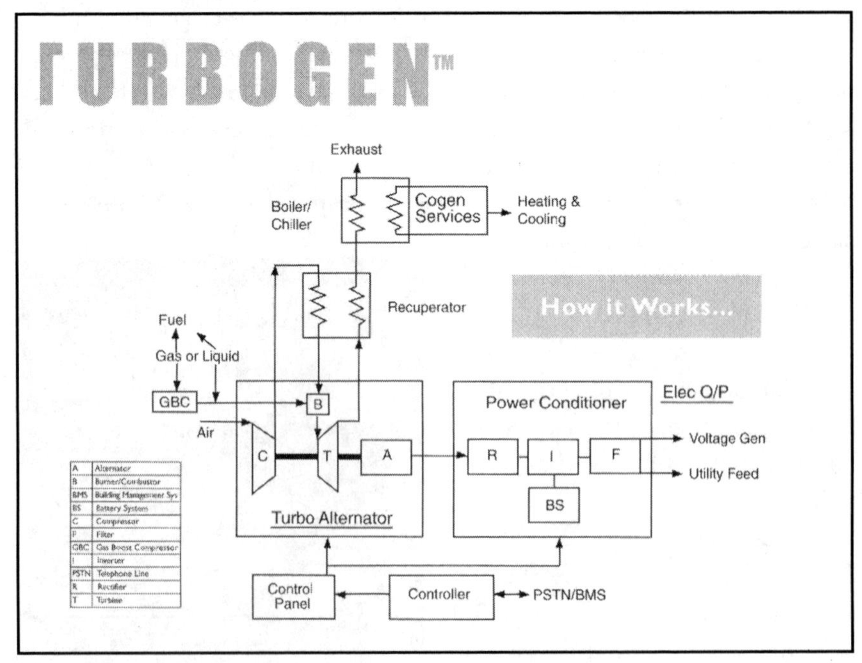

***Fig. 3-4** A drawing of the BPS System Arrangement.*

Capstone Turbine Corporation. Capstone Turbine Corporation (Capstone) is located in Woodland Hills, California. It was founded in 1988. According to Capstone, it is recognized in the industry for its research, development, and field applications of advanced turbine-driven generator technology.

Capstone's premiere product is the Capstone Microturbine. Its development began in 1993. Its technology is based on the same technology as a jet engine, except that it integrates Capstone's patented air bearings and proprietary software using state-of-the-art electronics. Capstone expects these innovations to create a versatile solution both for electric and thermal energy.

The Capstone Microturbine Model 330 is a fourth generation MTG. Many believe it reflects its maturity in terms of component arrangement, power electronics, control system, and overall packaging.

Figures 3-5a, and 3-5b show the Capstone 60kW, Capstone's newest microturbine.

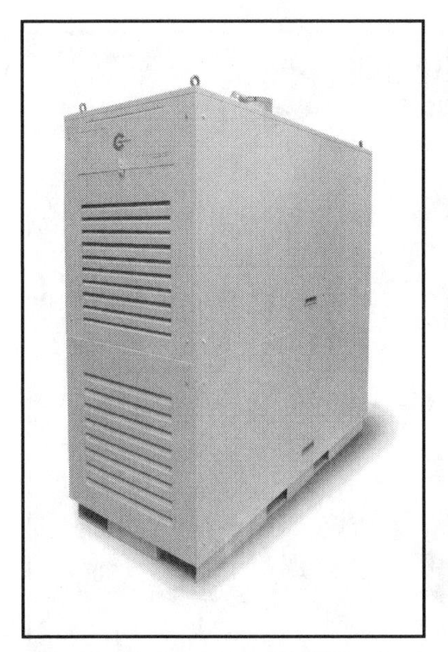

***Fig. 3-5a** Capstone 60 kW microturbine*

***Fig. 3-5b** Capstone 60 kW microturbine with panel removed.*

Elliott Energy Systems. Elliott Energy Systems (EES) is headquartered in Stuart, Florida. EES was formed in 1996 and is a subsidiary of the Elliott Turbo Machinery Company. It was founded in 1895 under the original name Chicago Boiler Company. Through the Elliott legacy, EES comes from a long history in steam turbines.

According to EES, it designs and manufactures MTGs from 35kW to 80kW, with plans to expand its product line to 200kW. In 1998, EES began to design its own recuperator in an effort to provide a more efficient and competitively priced unit. EES distributes and services its products through a worldwide network of independent distributors and integrators. The distribution channel will be enhanced in the future by packaging facilities in Japan and Eurpoe.

EES manufactures its Turbo Alternator in sizes of 35 kW, 45 kW, and 80 kW. Figure 3-6 shows EES' 80 kW Turbo Alternator set. Figure 3-7 lists the design features and specification for the Turbo Alternator 80 kW.

Fig. 3-6 *EES 80 kW Turbo Alternator set.*

Model TA 80R, 80 kW Recuperated Turbo Alternator ™ Set

Turbo Alternator™ Set Rating

Three Phase	60 Hz kW/KVA	50 Hz kW/KVA
Standby	80 / 80	80 / 80
Prime Power	80 / 80	80 / 80

Systems Rated at 59° F, Sea level

Quality Power Producing Equipment

is our business at Elliott Energy Systems, Inc.. Our power systems offer solutions to requirements for reliable, quality electrical power.

- 100% full load tested.
- Performance supported by prototype testing.
- 50 or 60 Hz operational.
- Standard Digital Control Panel meeting standards set by NFPA-110.
- IEEE 519 Compliant

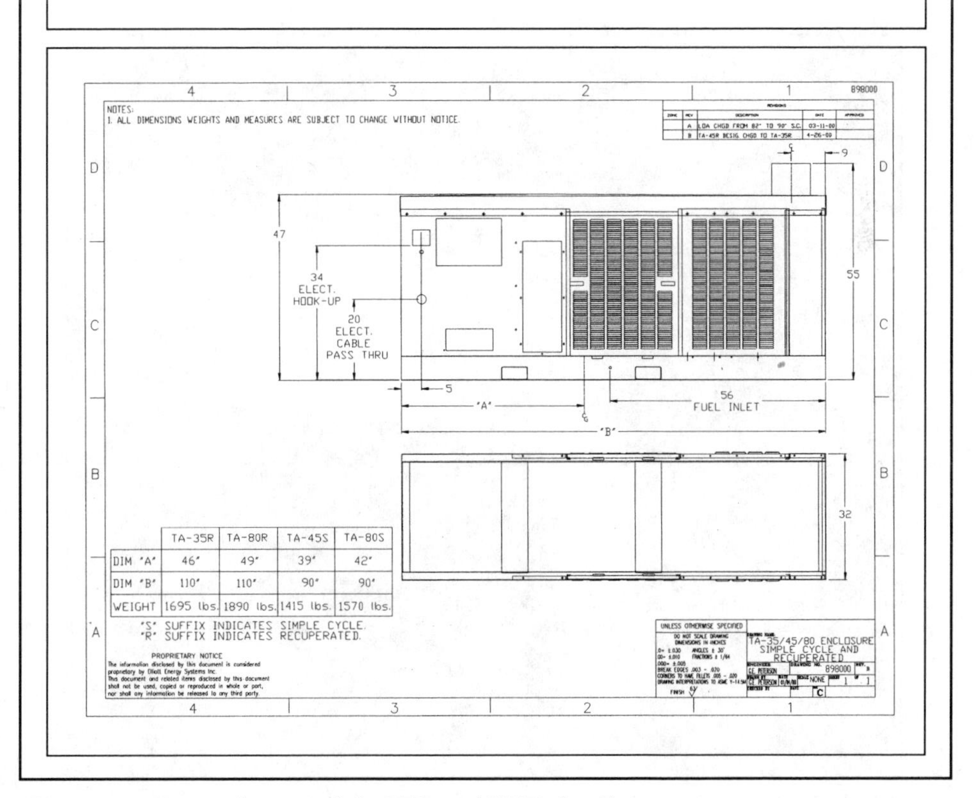

Figure 3-7 Design features of the EES 80 kW Turbo Alternator set.

Honeywell Power Systems. Honeywell Power Systems has its worldwide headquarters and manufacturing in Albuquerque, NM, with additional sites in Torrance, CA and Phoenix, AZ. It was originally created in 1994 as a part of Allied Signal as Allied Power Systems. Following the merger of Allied Signal and Honeywell, the merged company adopted the Honeywell name and Allied Power Systems became Honeywell Power Systems. Honeywell is being acquired by General Electric.

Honeywell offers the Parallon 75. According to Honeywell, it is developing and marketing its Parallon products for commercial, industrial, and transportation applications.

Figure 3-8 is a photo of the Parallon 75. Table 3-3 shows a specification for the Parallon 75.

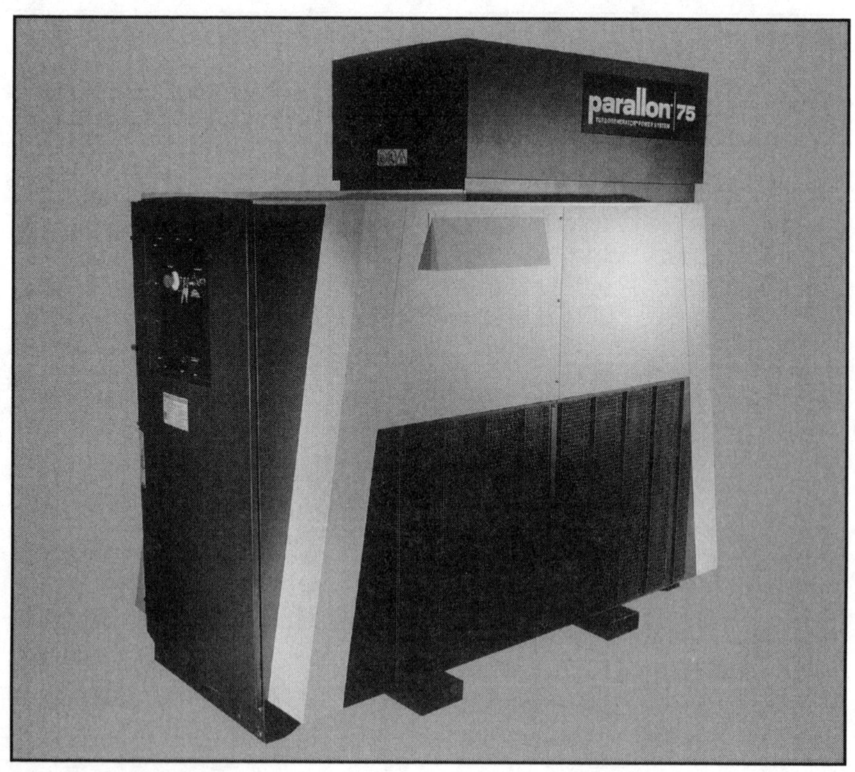

Fig. 3-8 *The Parallon 75.*

Quick Specifications

MAXIMUM POWER AT ISO CONDITIONS (59° F AND SEA LEVEL)	75 kW continuous rating.
THERMAL TO ELECTRICAL EFFICIENCY (INCLUDING AUXILIARIES, LESS GAS PUMP)	30% target, 28.5% target 27% guaranteed minimum at maximum power ISO conditions at minimum heat content of 19,500 BTU/LBM/LHV.
VOLTAGE OUTPUT	Options for 120/208, 120/240, 230/400, 240/415, 277/480, 360/600 all 3-phase or wire (with optional transformer), 50 to 60 Hz. Single phase operation must be balanced within 10%.
AVAILABILITY / UPTIME	>95%.
DIMENSIONS	Approximately 92" (2334 mm) L x 85" (2163 mm) H x 48" (1219 mm) W.
WEIGHT	Approximately 2850 lb (1295 kg), not including optional gas compressor, transformer and battery.
FUEL CONSUMPTION	1000cfh or 9.5 Therms per hour.
FUEL PRESSURE	75-85 psig [optional integral gas compressor for low pressure (>7 inches) available].
START UP	Normal start: 2.5 minutes.
NOX EMISSIONS AT 15% 0_2	50 ppm standard day.
NOISE	65 dBa at 10 meters, low frequency.
DESIGN LIFE	Following the recommended maintenance program, the unit is designed to operate a minimum of 40,000 hours. The useful life of the machine under normal operating conditions is expected to be 10 years.
WARRANTY SERVICE	One year from date of installation, not to exceed 18 months from purchase. Honeywell Inc. is the authorized warranty service provider in North America.
*Product and service descriptions and/or specifications are subject to change without notice or liability.	

Table 3-3 *Specifications for the Parallon 75.*

PowerWorks.

70 kW Microturbine Cogeneration System

Draft Product Specification

A rugged and long-life cogeneration system driven by a low-NOx, recuperated gas turbine engine

Key Features

- 70 kW at ISO Conditions
- Simple, "ruggedized" turbocharger-based design
- Industry's best recuperator:
  - Easily withstands harsh engine temperatures
  - No limit to number of engine start/stops
  - Allows microturbine to reach full power more quickly
- Low emissions that easily meet California emissions standards
- Integrated, variable output waste heat recovery system
- Natural gas, propane, or #2 diesel fuel (optional)
- Internal, 80,000 hour life fuel gas booster (optional)
- Proven oil-lubricated bearings
- Product life 80,000 hours
- Mean time between forced outage 8,000 hours

Target Specifications

Efficiency
Electrical: >27% HHV
>30% LHV
Overall: Up to 80% LHV

Emissions (Natural Gas)
NOx: <9 ppmv @15% O2
CO: <9 ppmv @15% O2

Physical
Size: 69L x 36W x 87H (in)
175L x 91W x 221H (cm)
Weight: 4100 (lbs), 1860 (kg)
Noise: 69dba @ 1m

Patented Recuperator
- Highest value component
- Critical to high efficiency
- Designed for 80,000 hour life

Two-Shaft Engine
- Reduces stress for longer life

Mechanical Shaft Output
- High efficiency platform

Proven Generator Technology
- Well understood by utilities, reduces utility barriers
- High efficiency
- Low cost
- Low emissions

***Fig. 3-9** Key features and specifications for the 70kW PowerWorks.*

Ingersoll-Rand Energy Systems. Ingersoll-Rand Energy Systems, formerly known as Northern Research Engineering Corporation, is located in Portsmouth, NH. It is an outgrowth of work done by Northern Research Engineering Corporation that was founded in 1956 for "scientific research and engineering." According to Ingersoll-Rand, PowerWorks, its family of microturbines, products are being developed for power generation, cogeneration, and other applications.

Ingersoll-Rand believes the added capabilities of two shafts will be advantageous to customers who can take advantage of application of a second shaft. Also it believes that the "free" power shaft reduces stress and prolongs engine life.

Figure 3-9 is a photo and description of the key features and the specifications for the PowerWorks.

Turbec AB. Turbec AB, owned 50/50 by ABB and Volvo, was founded in 1998. According to Turbec AB, its mission is to create a global business in microturbine products. The first stationary MTGs expected from Turbec are a 100 kW CHP followed by a 40 kW unit. The experience to develop the MTG came from the automotive side of Volvo and dates back to the early 1970s when its cars used the 40 kW size, and the 100 kW size was used on its trucks and buses.

Volvo joined with ABB to create Turbec to take advantage of the Volvo technology and manufacturing "know how" and ABB's strength and global market presence in the power generation business.

Figure 3-10 shows the Turbec engine design and Figure 3-11 is a photo of the T100. Turbec's website, turbec.com, lists the specifications for the T100.

Origin and Operation of MTGs

MTGs are small, high-speed power plants that usually include, on a single shaft, the turbine, compressor, and generator. Most MTGs designs use power electronics to convert high frequency generated power to standard frequencies and voltages and deliver power to a customer or a grid. They typically operate on natural gas but can also operate on propane. Some units run diesel and others on lower energy fuels such as gas produced from landfill or digester gas. MTGs originated through advances in three primary sectors –

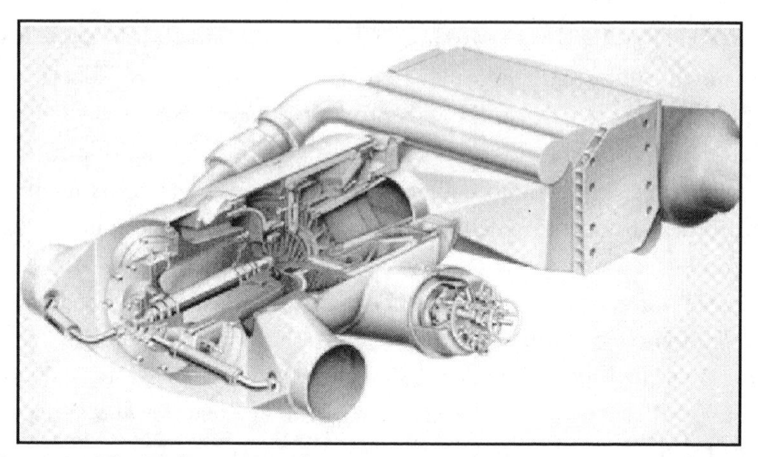

Fig. 3-10 *The Turbec engine design.*

Fig. 3-11 *The Turbec T100.*

transportation (turbochargers), aerospace (APUs), and the turbines and engines industry.

Most MTG designs have a high-speed (70,000 to 120,000 rpm) gas turbine engine driving a generator. MTGs come in a variety of maximum output sizes. The range is about 20 to 500 kW. Single-shaft MTG designs produce high frequency electric power, 1000s of Hz, which is converted to high-voltage DC and then inverted back to 60 Hz. The voltage produced by the inverter varies with the inverter design. For example, Capstone uses 277/480 (phase-to-neutral, phase-to-phase) and Honeywell uses 275 VAC phase-to-phase.

An MTG uses energy to produce electricity and usable byproducts such as heat. The MTG burns fuel in air, raising thermal energy of the air. The air's thermal energy is converted to kinetic energy in turbine nozzles. The high velocity air strikes and impels the turbine blades converting the air's kinetic energy to turbine mechanical energy.

The turbine's mechanical energy is sufficient to drive the compressor and generator. The air exhausting the turbine retains considerable thermal energy, which is available for use in a low temperature process.

The turbine, compressor, and generator are mounted on the shaft. The MTG is started by the generator spinning the shaft. Typically, for a single-shaft MTG grid-connected, this initial spinning is accomplished by motorizing the generator from the grid. After it reaches sufficient speed and air flow through the compressor, fuel is injected into the combustion chamber and ignited. The energy of combustion sustains the flame, and the igniter is turned off. After a short warm-up and ramp-up to operating speed, the turbine is ready to load for sustained operation.

A distinguishing characteristic of some MTG designs is the ability to start without being connected to the grid. This is called "black start" capability. An outside power source is needed to start the MTG – usually batteries – which can be used to motorize the generator for starting the turbine.

During operation, engine air is drawn into the unit by the compressor and discharged through a recuperator, where the temperature is increased by recovering some thermal energy from turbine exhaust gas. Recuperation is only practical for single-stage compressor gas turbines. The temperature gained in the recuperator reduces the fuel required to be burned to achieve the design turbine air inlet temperature (800 – 1000 degrees C). This heated air flows into the combustor where it is mixed with fuel, ignited, and burned.

The combusted gas passes through the turbine nozzle and turbine wheel. The turbine wheel is usually operating at high speed. The speed varies among manufacturers, generally ranging between 70,000 rpm and 120,000 rpm. For example, the Honeywell operates at 76,000 rpm, the Capstone at 96,000 rpm, and the Bowman at over 100,000 rpm with the turbine speed being model-dependent.

As the hot air expands through the turbine, it converts its thermal energy into the rotating mechanical energy of the turbine, which drives the compressor and generator. The gas exhausting from the turbine is directed back through the recuperator. The recuperator is a heat exchanger that delivers some of the exhaust gas energy to preheat air entering the combustor. A heat recovery unit can be added to recover residual waste heat in a combined heat and power (CHP) application, such as hot water.

BACKGROUND

Previous R&D

MTG technology is coming from research and development (R&D) in three industries — transportation, aerospace, and turbines/engines. In the late 1980s, major manufacturing breakthroughs in transportation resulted in the ability to mass-produce small turbochargers cheaply and of sufficient quality for use in stationary power applications. Turbochargers are small and simple with few moving parts.

Another important technology innovation borrowed from aerospace is the creation of air bearings used in some MTG designs. Air bearings eliminate the need for liquid lubrication and thereby make maintenance cleaner, simpler, and easier to handle.

Likewise, R&D in gas engines/turbines has improved the efficiency and overall operations so as to provide improved components, such as recuperators for MTGs.

Market Size

Manufacturing capabilities offer the potential of low cost products if a market large enough for high-volume manufacturing of MTGs can be secured. The key to success for MTGs will be in the economies of manufacturing. The could result in MTGs priced at no more than $500 per kW by 2005. For the MTG industry to become sustainable, many believe this cost target range must be met. However, a recent study shows industry experts believe that a sizable market can be secured at or below this price.

Operating Characteristics

Simple Design, Advanced Electronics

Except for its power electronics – which are complex – the overall MTG relies on a simple design with few parts. Of those parts, few are moving parts, thereby enhancing the simplicity of the design and its operation and potentially increasing its overall reliability and maintainability. And though power electronics are complex, it is a technology that is quickly advancing. Most experts believe that inverters will become more reliable and less expensive.

In essence, the MTG is a small combustion turbine that has an integral high-speed generator. The generator is at one end of the shaft next to the compressor, followed by the turbine.

Innovative Characteristics

Some MTGs use air bearings. Air bearings require no oil lubrication. Use of air bearings versus an oil lubricating system eliminates oil filter and oil replacement maintenance, and incorporates oil cooler in the design.

Operating Components

Turbines—The Core of the Power-producing Section

There are two types of MTG turbine designs — single shaft and dual shaft. In this section, the two different design types will be described and the attendant components resulting from each design will be explained. In a later section, the operational aspects will be compared along with expected advantages and disadvantages.

Currently, all MTG manufacturers other than Ingersoll-Rand use a single shaft design. For illustration of a representative single shaft design, the Honeywell Parallon 75 system configuration is shown in Figure 3-12. The single shaft operates at high speed. Depending on the manufacturer and the model of the MTG, speeds vary from 70,000 to 120,000 rpm. An inverter is required to provide power at 60 Hz.

For comparison, Ingersoll-Rand's PowerWorks dual-shaft configuration is shown in Figure 3-13. Through reduction gear, the generator speed is reduced to 3,600 rpm. No inverter is required to produce power at 60 Hz. With an induction generator, this design is straightforward, and the grid will ensure speed is maintained at 60 Hz. With a synchronous generator (grid independent), control of power turbine speed is essential to maintain 60 Hz. This can be difficult to operate on a stand-alone basis. Connected to the grid, the grid frequency will control speed, but the speed has to be controlled while paralleling with the grid.

MTGs with a single-shaft design mount the compressor, turbine, and the electrical generator on the shaft. This type of design has only one major moving part.

A dual-shaft design requires the gearbox and associated moving parts necessary for the gearbox as well as the gasifier turbine/compressor, and the power turbine, which represents the two separate turbines.

Alternator/Generator — sustains combustion and net power export. Alternator designs vary among manufacturers. On the single-shaft machines, it is mounted on the shaft and turns at high speed. At start-up, the generator is used as a starter motor. It drives the shaft-

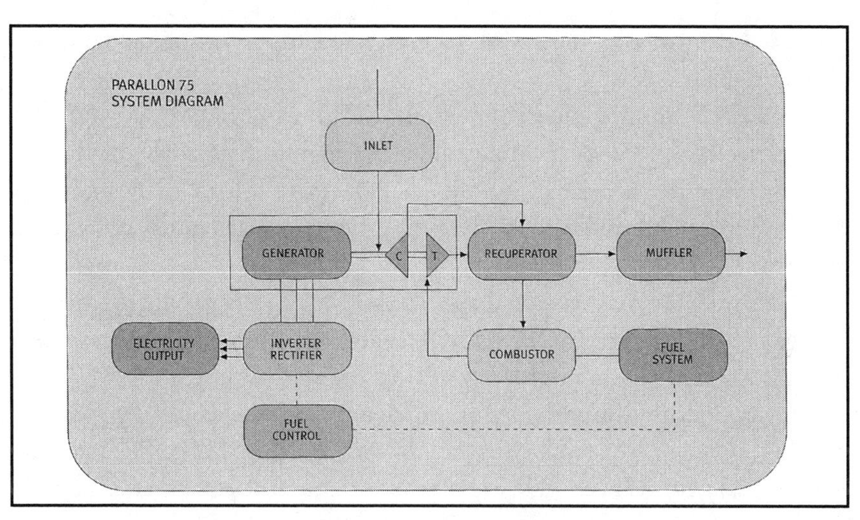

Fig. 3-12 *The system configuration of the Parallon 75 system illustrates a single-shaft design.*

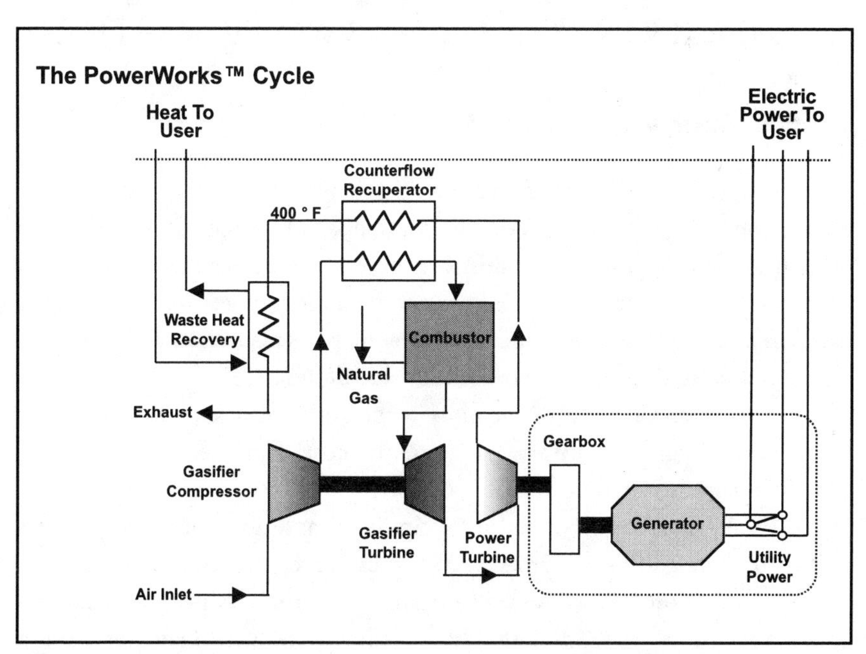

Fig. 3-13 *Ingersoll-Rand Energy Systems' PowerWorks incorporates a dual-shaft configuration.*

mounted unit to rotation speed necessary to support sustained combustion and export power.

Power Electronics—key component. MTG manufacturers use different power electronics. Some manufacturers use highly integrated power electronics that are proprietary to the manufacturer. Single-shaft machines require power electronics to convert high frequency generated power to a standard of 50 or 60 Hz power.

Control Systems—essential to success. MTGs rely on microprocessor based control systems. Similar to inverters and other power electronics, these systems are expected to improve in quality and decline in price.

Recuperators—important for efficiency. Unrecuperated MTGs are rated at about one-half the efficiency of recuperated MTGs. Recuperators are necessary for MTGs. MTGs generally exhibit efficiency that at best are still below 30%, even with a recuperator.

Comparison of Design Configurations

Single Shaft vs. Dual Shaft

Of the six manufacturers cited in this chapter, all but one use a single-shaft design. Ingersoll-Rand alone uses a dual-shaft design. Each design has certain capabilities and limitations. Customers will determine which design best suits their requirements. Only one shaft means no gearbox reduction is necessary. It also eliminates numerous moving parts associated with the gearbox and greatly simplifies the overall turbine design.

However, an offset to the gearbox is the addition of an inverter for the single-shaft design. An inverter adds to the cost, complication, and complexity of sophisticated power electronics.

A single-shaft design offers simplicity whereas a dual-shaft design offers ability to run additional processes such as air compressors and chillers. Running additional processes will be important to certain customers who can make use of an additional shaft. Other customers may not have a requirement for additional processes and will want the simplicity of a single shaft for electricity only or CHP.

Low Pressure Advantages/Limitations

Low pressure MTGs rely on an integrated natural gas compressor to boost the pressure. For example, Capstone requires 50 psig, Honeywell needs 75 psig, and Bowman requires 87 psig. The pressure is necessary to flow fuel into the combustor. The gas has to flow through a control valve, which regulates the flow rate. This valve will have a pressure drop across it depending upon the design of the valve. The supplied pressure to the inlet valve must maintain the necessary pressure level. For example, if the valve experiences a 20 psig drop, and the combustor is operating at 30 psig, then the gas pressure must be 30 + 20 = 50 psig. This points out a feature of the Capstone gas compressor—it uses a variable speed drive; therefore, its discharge pressure is slightly higher than the combustor's pressure. It controls the rate of fuel flow by varying speed and therefore no fuel flow control valve is required.

The gas compressor is needed because the natural gas infrastructure generally does not deliver gas at such pressures. However, the disadvantage of adding the compressor is that the compressor adds a parasitic load to the MTG. The additional parasitic load lowers the overall efficiency of the MTG and reduces the total net power output of the MTG. The addition of the compressor results in a greater ability to site the MTG at almost any allowable location on the natural gas pipeline. Capstone, equipped with a gas compressor, claims it requires at least 5 psig from the gas pipeline while Honeywell claims it needs only 0.3 psig to operate. At some locations on the gas pipeline, the pressure is as low as 2-3 psig. It will be interesting to see how well MTGs perform at some of these ultra low-pressure locations.

Low Temperature Advantages/Limitations

The advantage of low temperature operation of MTGs is the ability to use conventional materials such as steel, instead of exotic metals or ceramics, for the manufacture of MTGs. Such materials can be procured in bulk, at standard prices, with no special preparations. High temperatures require existing exotic (expensive) materials, such as specialty alloys or the development of new materials, such as ceramics (even more expensive).

Performance Criteria

Efficiency to Improve

MTG efficiency initially will be low compared to new large combined cycle (100+ MW) gas-fired plants. In fact, for most MTG models, the efficiency of a new large combined cycle gas-fired plant will be almost twice that of the MTG. However, new, large combined-cycle, gas-fired central plant efficiency will be offset somewhat by the added economic burden of building large, long-distance high voltage transmission lines. These lines are needed to move power from remote locations of central plants to the ultimate customer site and the transmission and distribution line losses related to transporting the power over long distances.

However, MTG efficiency should also improve as ceramic components are developed that allow the MTG to operate at higher temperatures. Operation at higher temperatures results in increased thermodynamic efficiency. In operations at high temperature, more fuel is burned to achieve the higher temperature, but even greater net work is produced and overall high temperature operation results in greater overall efficiency. Additionally, MTG efficiency should also improve as better recuperators are designed and built that recover heat more efficiently.

Noise to be attenuated or diffused. Most MTGs claim to operate at 65 to 70 dBA. This is about the same noise level as a large outdoor air conditioning unit. However, the high-speed turbine has a high frequency pitch. Some people find the high frequency sound annoying almost to the point of discomfort. This level of noise and the high frequency sound will limit potential installation sites unless special measures are taken to reduce the level of noise. Most of the manufacturers expect to bring noise down to 50 to 55 dBA over the next few years.

Emissions to differentiate MTGs from competitors. Some manufacturers expect to have historically low NO_x due to their development of low NO_x combustors. These manufacturers are advertising no more than 9 ppm for NO_x.

However, although some MTGs offer low NO_x emissions, MTGs are not highly efficient in this regard. Most are below 30% efficiency at the standard temperature and altitude whereas new large combined cycle gas-fired central plants are almost twice as efficient. Low efficiency results in having to burn more fuel and sustain more combustion than a power-producing process with higher efficiency. Combusting natural gas results in CO_2, a "greenhouse" gas. So by increasing the efficiency, the amount of CO_2 released during combustion is decreased.

Power Quality is important. MTGs will need to provide power quality in accordance with IEEE Standard 519. This standard is a function of size of load and location of the load on the distribution or transmission grid. If the MTG does not meet this power quality standard, it may not perform as desired by the customer.

Energy Conversion

Gas turbines convert the chemical energy of the fuel to the mechanical energy of rotating machinery that drives the generator to produce electrical power. Exhaust heat can be captured to apply thermal energy for use, such as hot water.

A compressor is required to compress air. The compressor increases the pressure three to four times atmospheric (about four times for most models of MTGs). The compressed air is then pushed through the remaining components of the MTG. These include the combustor, turbine, and recuperator.

The air is heated by a combustor burning fuel. Air pre-heating occurs in the recuperator prior to combustion.

Air is expanded through nozzles and the turbine. Doing so converts thermal energy to mechanical energy and lowers air temperature and pressure.

Air is cooled by the recuperator that delivers exhaust energy to compressed inlet air and by a heat recovery system for CHP models.

The mechanical energy of the rotating turbine drives the compressor and generator. The generator produces electrical power.

Potential Applications

End-use customers will want to gain the advantage of on-site generation for as many applications as economically valuable. The following applications for MTGs will be briefly discussed:

- Stand-alone
- Peak shaving
- Standby and Emergency
- Energy portfolio management
- Aggregated dispatch
- Resource recovery

Stand-alone power applications usually occur in remote locations where the electrical grid is not available. It can include applications such as construction sites, oil fields, and new developmental or isolated operations.

Peak shaving can reduce monthly demand charges and increase year-round savings. By lowering the annual peak demand, customers may be able to reduce the monthly demand charge. In a simple example, reducing the annual peak from 100 kW to 80 kW could reduce the demand charge by 20% depending upon the structure of the applicable utility tariff. Actual savings will depend upon how demand charges are structured on the applicable utility system tariff.

Standby and Emergency Applications provide back-up power when the grid is down.

Energy portfolio management makes use of the MTG as a physical hedge when price spikes are passed through directly to customers.

Aggregated dispatch for utility and RTO ancillary services can be valuable if enough MTGs can be made available to the utility or the RTO to provide capacity, energy, and other ancillary services during periods of shortage.

Resource recovery is an extremely important application for MTGs. MTGs that can burn low energy gases and sour gases as fuels for power can be highly economically valuable. Otherwise, the low energy content fuel is wasted and cannot be turned into electricity.

The Future

The U.S. Department of Energy (DOE) has identified turbines as one of the 27 critical technologies for U.S. security and prosperity. As such the DOE offers funding for research, development, and demonstration (RD&D) for MTG and MTG component development, such as recuperators and for new materials, such as ceramics.

The DOE uses three important criteria to award funding. These criteria are reduction of energy consumption, improvement in environmental conditions, such as emissions, and improvement in the overall economics of the technology needed to increase the advantage of U.S. manufacturers to sell their products worldwide.

Southern California Edison's Testing Program

SCE conducts a unique microturbine testing program for the DOE, the California Energy Commission, and EPRI. Figures 3-14a and 3-14b are the DOE's description of the program. The testing is carried out at the host site. The selected host site is at the Combustion Laboratory at the University of California, Irvine (UCI) in Irvine, CA. UCI was chosen for its robust advanced power program that features education and research for energy technologies. The program relies on the National Fuel Cell Research center, the world-renowned Combustion Laboratory, and UCI's Distributed/Dispersed Energy Technologies program and demonstration facilities, including the development of a connectivity laboratory, all housed at UCI.

The MTG testing program began in 1996. It has tested MTGs from Capstone Turbine Corporation and Bowman Power Systems. In 2000, the program purchased and began testing a Honeywell 75 kW and a Capstone low-pressure 30 kW. An Elliott 80 kW is in installation to begin testing. Ingersoll-Rand and Turbec are not presently in commercial production so no MTGs are available for purchase and testing.

The program tests for machine performance. It tests each MTG's performance against its manufacturer's performance claims for efficiency, emissions, and noise. Also MTGs are tested against applicable industry standards,

INDUSTRIAL POWER GENERATION PROGRAM

Project Fact Sheet

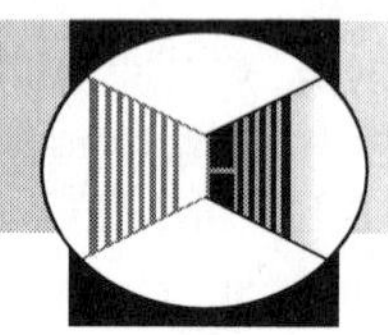

ADVANCED MICROTURBINES

BENEFITS

Microturbines offer a number of potential advantages compared with other technologies used for small-scale power applications, including:

- higher reliability due to the small number of moving parts
- simplified installation
- low maintenance requirements
- compact size
- light weight
- acceptable noise levels
- acceptable efficiency
- fueled by domestic natural gas resource with expanded fuel flexibility
- competitive cost when built in quantity
- low emissions
- high temperature exhaust for heat recovery
- acceptable power quality

APPLICATIONS

Due to its potential advantages, microturbines can be used as a distributed generation resource for power producers and consumers, including industrial, commercial, and, in the future, even residential users of electricity. Significant opportunities exist in four key applications:

- Traditional Cogeneration
- Generation Using Waste and Biofuels
- Backup Power
- Remote Power for those with "black start" Capability
- Peak Shaving

MICROTURBINE GENERATOR TESTING AND EVALUATION

The U.S. Department of Energy (DOE), in cooperation with the gas turbine industry, has initiated the Industrial Power Generation (IPG) Program to develop advanced engines and technology to meet industrial power generation needs. The IPG Program will help contribute to the development of ultra-high efficiency engine systems for industrial markets.

One of the supporting elements of the IPG Program is the testing and evaluation of microturbines. Southern California Edison in partnership with University of California at Irvine, is leading the project through the Microturbine Generator Development and Assessment Program.

Microturbines are primarily fueled with natural gas and can generate anywhere from 25 kilowatts to 200 kilowatts of electricity. By locating power generation facilities such as microturbines close to the loads they are serving, the need for transmission and distribution equipment may be reduced, and it may be possible to provide pinpoint control and enhance reliability. Along with other small-scale distributed generation options, microturbines could expand the range of technology options for industrial consumers in competitive power markets.

MICROTURBINES

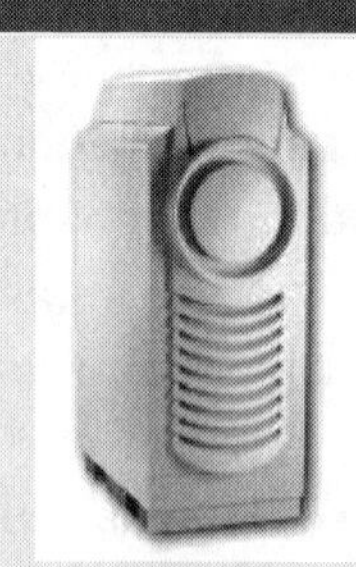

Capstone Turbine Corporation 28 kW microturbine (left) and Bowman Power Systems 35 kW microturbine (right) are being evaluated at the MTG test center.

OFFICE OF INDUSTRIAL TECHNOLOGIES
ENERGY EFFICIENCY AND RENEWABLE ENERGY • U.S. DEPARTMENT OF ENERGY

Fig. 3-14a Description of the MTG testing program.

Project Description

Goal: The primary goal of the project is to establish the technology baseline for microturbines and guide future Research and Development needs. Southern California Edison in conjunction with the IPG Program and industry partners will perform the following to achieve this goal:

- Install, operate, maintain and test microturbine generators at the Microturbine Generator (MTG) test center at University of California at Irvine (UCI).
- Assess and report the performance characteristics of installed microturbines including efficiency, reliability, endurance, emissions, ease of maintenance, and economy of operation.

Progress and Milestones

- The UCI MTG test center infrastructure was completed in 1998. It includes adequate compression of MTGs with compressors. It includes electrical and gas systems including the ability to use "lower than pipeline quality" natural gas for testing and provides four skids for simultaneously testing up to four MTGs.
- A remote data acquisition system (DAS) has been implemented.
- In 1999, Southern California Edison is testing two Bowman MTGs (35 kW and 60 kW) and a Capstone MTG (28kW).
- Southern California Edison plans to test a Bowman (45 kW), an AlliedSignal (75 kW), and two Elliott Energy Systems' MTGs (80 kW and 200 kW).

MICROTURBINES

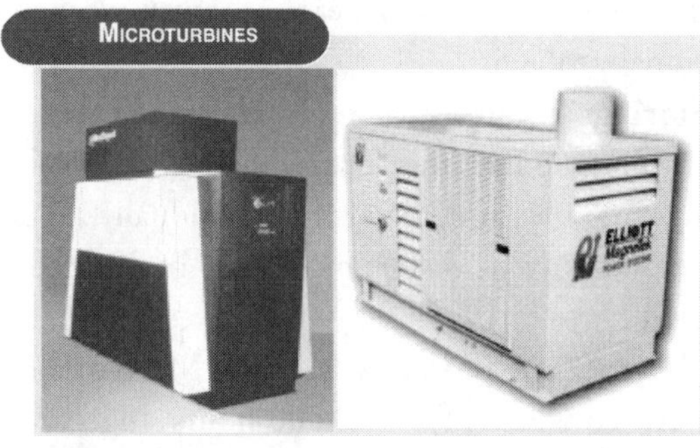

AlliedSignal Power Systems Incorporated 75 kW microturbine (left) and Elliott Energy Systems 45 kW microturbine (right) are being evaluated at the MTG test center.

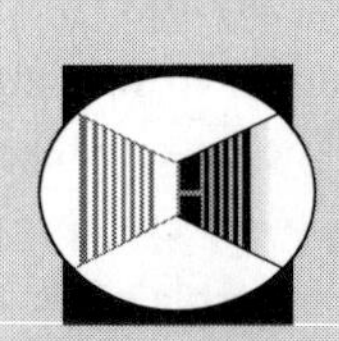

PROJECT PARTNERS

Southern California Edison
Rosemead, CA

AlliedSignal Power Systems Incorporated
Albuquerque, NM

Bowman Power Systems
Southhampton, England

Capstone Turbine Corporation
Woodland Hills, CA

Elliott Energy Systems
Stuart, FL

University of California at Irvine
Irvine, CA

FOR ADDITIONAL INFORMATION, PLEASE CONTACT:

Stephanie Hamilton
Southern California Edison
Phone: (626) 815-0514
Fax: (626) 334-0793
Shamilto@edisontec.com

Doug Hooker
DOE Golden Field Office
Phone: (303) 275-4780
Fax: (303) 275-4753
doug_hooker@nrel.gov

Debbie Haught
Office of Industrial Technologies
Phone: (202) 586-2211
Fax: (202) 586-1658
debbie.haught@ee.doe.gov

Visit the Program's home page at
www.ms.ornl.gov/ats/

Please send any comments, questions, or suggestions to
webmaster.oit@ee.doe.gov.

Visit our home page at
www.oit.doe.gov
Office of Industrial Technologies
Energy Efficiency
and Renewable Energy
U.S. Department of Energy
Washington, D.C. 20585

October 1999

Fig. 3-14b Description of the MTG testing program (continued).

such as power quality and/or local requirements, such the South Coast Air Quality Management District's (SCAQMD) air quality emission standards for NO_x and CO_2. Also, qualitatively assessed are the ease of MTG installation and startup, maintenance and operation, and overall machine performance. The on-site testing crew maintains a daily log of testing activities. The daily log is kept to ensure the integrity of the testing results and to record events to explain the data captured or why data is not being captured.

Testing results include the following:

Starts/stops: ideally the number of starts and stops are equal for each testing session when the MTG is started up and intentionally shut down. A large difference in the number of starts and stops indicates that the machine is experiencing problems.

Overall unit efficiency and net power output: based on actual conditions the machine should provide a level of efficiency, within a small test-specified tolerance, as predicted by, and consistent with, the derating curves provided by the manufacturer.

Operability: subjective assessment of the machine's ease of operation, performance reliability and consistency, and its ease of return to operations after experiencing operational problems.

Emission level monitoring: within a small test-specified tolerance, emissions are expected to be within manufacturer's claims for NO_x and CO_2.

Power quality monitoring: measures distortion individually for current and voltage. Both voltage and current distortion should be in accordance with the IEEE 519 standard.

Endurance testing: is a measure of longevity of the MTG's wheel life. Most MTGs have an advertised wheel life of 40,000 hours. Ingersoll-Rand Energy Systems advertises 80,000 hours of wheel life.

As a part of the testing program, facilities at UCI have been established that provide a "level testing field" for all MTGs. MTGs are equipped with data acquisition equipment to ensure that data is electronically captured on a real-time basis. A veteran on-site two-person testing crew also reviews the electronic data capture with manual measurements and calculations, as necessary, to ensure that electronic capture is consistent with physical experience.

The testing crew interacts with technical staffs from the manufacturers. An essential part of the testing program is to provide written feedback to each manufacturer, on an individual basis, about the results of the testing program for each manufacturer's individual MTG. The testing crew offers suggestions for consideration by the manufacturer for future product enhancements.

Another important value of the testing program is that it provides independent third-party information for the public on the performance of the MTGs on a consistent basis under actual operating conditions. "Lessons learned" offers expert advice on operating experience and observations that can be used by the public to consider how best to use MTGs under actual operating conditions.

DOE Summit

To advance MTG development, the DOE sponsored the Microturbine Technology Summit in December 1998. The summit was intended to surface issues so that a thoughtful roadmap would emerge for focused and results-oriented research, development, and demonstration (RD&D). The DOE successfully obtained valuable ideas and comments from a group of industry stakeholders to help in its efforts to develop a RD&D program for microturbines. Both policy- and market-related issues were necessarily a major part of the discussions.

The Summit identified that the market for MTGs is potentially quite large but the alternatives that are competing to serve industrial power needs will be hard to beat with today's existing MTG technologies. The favorable attributes of fuel cells – another emerging technology – put lots of pressure on further development, improvement, and enhancement of MTGs.

The major findings of the DOE Microturbine Summit are as follows:

Achieving the goal of increasing the overall efficiency of microturbines to 40% or greater could boost the appeal of microturbines substantially compared with competing technologies.

A number of barriers affects the development of markets for small-scale power plants, including microturbines, not the least of which is uncertainty about the future of the structure of electric power markets.

A particular issue is the interconnection of distributed generation technologies, including microturbines, with the utility grid. Interconnection specifications are not standardized and vary by utility system across the United States.

A focused RD&D program can be a great help in improving the prospects for microturbines.

Lower-cost, more efficient microturbines with known performance and proven reliability are needed.

RD&D to lower cost and increase the reliability of equipment for fuel processing, gas compression, recuperation, and power electronics is also important.

Development of advanced materials that are less costly, more durable, and more capable of operating efficiently at higher temperatures could be one of the keys to making substantial improvements in the thermal efficiency and environmental performance of microturbines.

As a result of the Summit, DOE and others have developed funding solicitations to provide for future RD&D funding focused on the findings above.

If the findings of the Summit are pursued, MTGs can be expected to increase efficiency through improved materials, such as ceramics, and improved components, such as more efficient recuperators and more advanced power electronics.

Expanding Niches Through Applications

MTG manufacturers during the next three years will add features targeted for expanding their entry niche markets. Expected niche markets are those that value reliability.

The next three years will focus on enhancing the MTG's capability in key areas that make it a broader based product. These capabilities include the following.

Plug & Play enhancements will be added to increase the user-friendliness of the MTG and its capabilities. Such ease of use and expanded capabilities will be advantageous for small customers who do not employ or expect to employ highly technical staff. These customers will require that the MTG be installed simply and operated unattended. The MTG must be smart enough to trouble-shoot problems and able to configure itself given the customer's physical and operating requirements and overall constraints. It should be able to advise of future maintenance, such as cleaning filters, replacement parts, at routine intervals. Future Plug and Play will have "smart" operations to know when to import or export power with the grid to minimize the customer's overall power costs. As time-of-use rates become widespread, MTG may be able to provide ancillary services.

Fuel flexibility with dual fuel capability. MTGs will need to operate efficiently on a variety of fuels, including natural gas, diesel, propane, and digester gas. Most of the manufacturers have realized the value of multiple fuel operations and have designed, or are designing future models that can operate on a variety of fuels. Additionally, the MTG will need to have the capability to switch between fuel types so as to provide back-up fueling capability. Ideally, these capabilities will be provided transparently to the customer requiring only simple modifications, if any.

Tight, seamless integration to the grid will be important to the MTG's economics for customers. MTG manufacturers are working with software/firmware providers to provide communications and controls that easily provide the ability to aggregate and centrally dispatch many dispersed MTGs, if used as standby, and other standby distributed generation technologies, such as storage. Small generators located in constrained parts of the grid can be dispatched and bring needed capacity during peak demand periods when spot prices can soar; however, communications, controls, and other necessary hardware and software must be installed to ensure proper and reliable operations.

Environmental issues related to MTGs include emissions and noise. MTGs are expected to be low in NO_x. Although not all MTGs have specifications, that if met, would meet the NO_x emission level set by the SCAQMD, many in industry believe that the SCAQMD's level of less than

9 ppm NO_x, corrected to 15% O_2, will become the industry goal. MTGs will continue to push down the level of NO_x. Meanwhile, the efficiency of MTGs compared to the efficiency of new, large combined-cycle, gas-fired central plants makes reducing MTG greenhouse emissions through increased efficiency an important goal. Most MTG manufacturers claim noise levels in the 65 – 70 dBA range at 10 meters. In certain urban locations, this level will need to decline to 45 – 60 dBA. Also, the high frequency pitch from the high-speed turbine will require significant attenuation in some locations and some applications.

The Next Five to Seven Years

During the next five to seven years, manufacturers will expand niches and attack similar niche opportunities. Initial niches for MTGs are commercial customers who value increased reliability due to significant costs related to spoilage or lost business. MTG manufacturers should look for the same type of customer in the industrial sector. To get at this larger-size customer, MTGs will need to be ganged up into multiple-unit packages.

Another attractive niche for MTGs is the customer who uses lots of energy in production processes and wants to benefit from managing energy price volatility. As electric industry deregulation continues, rates will move toward "time of use." Under time-of-use pricing, electricity is priced and sold in discrete blocks of time. During peak periods of the day, prices can soar. In this instance, the MTG can provide a physical hedge.

Survival

Surviving into the next decade means there must be major improvements in overall product robustness and performance so as to grow into broad applications and secure market acceptance. Improvements in the next decade to a large extent will track the DOE's criteria for improved efficiency and lower emissions.

The most challenging and important issue will be to improve the efficiency of the MTG to 40+% without raising the capital price, the cost of maintenance, or complicating the operation of the machine. Without this efficiency improvement, MTGs will not compete well with emerging fuel cells technology, especially given the added environmental benefits of fuel cells in terms of no emissions and no noise. To achieve the efficiency gains, operating pressures and/or temperatures will need to be increased.

End Notes

1. The Role of Distributed Generation in Competitive Energy Markets, Distributed Generation Forum, 1999.
2. Building Operating Management, March, 2000, page 12, "Outlook, Minipower Plants: Microturbines Draw Interest."
3. "Distributed Generation: Understanding the Economics," An Arthur D. Little White Paper, 1999.
4. "Don't underestimate challenges of bringing gas to powerplants," POWER, January/February 2000, p. 27-30.
5. Hamilton, Stephanie L. "Microturbines poised to go commercial," Modern Power Systems, September 1999.
6. "Advanced Microturbines," DOE's Office of Industrial Technologies, Energy Efficiency and Renewable Energy, Project Fact Sheet.
7. Watts, James H. "Microturbines: A New Class of Gas Turbine Engine," Global Gas Turbine News, Volume 39: 1999, No 1.
8. de Rouffignac, Ann. "Backing Up the Grid with Microturbines," RDI Energy Insight, December 3, 1999.
9. Wheat, Doug. "Distributed gen enhances the grid, but can't beat central power," POWER, November/December 1999.
10. Swanekamp, Robert. "Distributed generation seeks market niches," POWER, November/December 1999.
11. Hamilton, Stephanie L. "The Buzz is from the Microturbine Generators," Deregulation Watch, 7.31.99, Vol. 2, No. 14.
12. Hamilton, Stephanie L. and Steven Taub, "Microturbine Generator Test Program Report," Cambridge Energy Research Associates Forum Report, December 1999.

13. "Distributed and Dispersed Energy Resources, A Paradigm Shift," NFCRC Journal, July/August/September 1998, Volume 1, Issue 3.
14. Hamilton, Stephanie L. "Microturbine Generator Project," Report to the California Energy Commission, September 1999.
15. Bahl, Prem K. and Stephanie L. Hamilton, "Microturbines Under the Microscope," POWER-GEN International Conference, New Orleans, LA, November30-December 2, 1999.
16. Zimmer, Michael J. "Distributed generation offers T&D cost management," Electric Light & Power, February 2000,Volume 78, Number 2.
17. "Hybrid system offers new on-site generation," Electric Light & Power, February 2000, Volume 78, Number 2.
18. Cibulka, Lloyd. "Distributed generation advances against industry barriers," Electric Light & Power, March 2000, Volume 78, Number 3.
19. "Elliott at 100+," Turbomachinery International, September/October 1999.
20. Crawford, Mark. "New Microturbines Offer Viable Power Alternatives", Energy Daily, September 2, 1999.
21. Betz, Kenneth W. "Distributed Generation Poised to Win Over End Users," Energy Users News, April 2000, Volume 25, Number 4.
22. Lihach, Nadine. "The Distributed Generation Puzzle: Piecing It Together," Power Engineering, April, 2000.
23. "Microturbine market holds promise and uncertainty," Electric Light & Power, May 2000, Volume 78, Number 5.

STEPHANIE L. HAMILTON BIOGRAPHY

Stephanie L. Hamilton is a recognized expert in distributed generation, writing and speaking internationally and specializing in micro turbine generators.

She directs a micro turbine generator testing and assessment program for a major utility and works closely with governmental agencies, industry organizations, and universities.

Ms. Hamilton has worked for several natural gas and electric utilities, with responsibility for hydroelectric and thermal. She has a B.S. in Mechanical Engineering and an M.S. in Management.

4 Internal Combustion Engines And Portable Power

Many of today's installed on-site generation facilities actually run on internal combustion engines that are relatively similar to those found in cars and trucks. These technologies are reliable and relatively inexpensive, although in power generation uses they are somewhat less efficient than some other, more expensive technologies. Because these systems are less expensive to install, but more expensive to run due to their fuel expenses, they are often used as backup systems to ensure power reliability at hospitals, hotels, and manufacturing plants where an outage would be expensive or dangerous.

Internal combustion engines can run on natural gas, gasoline, diesel fuel, petroleum, alcohol, or other liquid fuels that have been converted to gases. Combustion in an engine happens in one of two ways. In the first, fuel in a gaseous form is mixed with air and compressed in a cylinder to such a high pressure that its temperature will ignite the fuel without the aid of a spark. Once the high pressure is attained, fuel is sprayed into the cylinder with the heated air.

Like automobile engines, engines used to generate electricity can have a varying number of cylinders. The number of cylinders is based on two requirements – the smoothness of the power supply needed and the total amount of power needed. Mechanical restraints and cost differences generally limit the numbers of cylinders used and how they are arranged. Space limitations are also often a factor in engine selection as well.

Engines that produce larger quantities of power may have larger cylinders and the firing of the engine may take place on both sides of the piston. In engines in which air is compressed separately, the fuel may be vaporized by passing through a nozzle under pressure or the liquid fuel may be pumped under pressure as a thin stream. If air needs to be mixed with the fuel before it enters the cylinders, this is done by using a carburetor or air injector.

Except for small gasoline-powered engines that are cooled by an air stream flowing past the engine, water or liquid chemicals cool practically all other internal combustion engines.

Engines are rated in terms of horsepower. One horsepower is the power required to lift 550 pounds one foot in one second. Engines are usually rated at their maximum horsepower and speed. In electric power generation, engines and generators are often packaged together in one modular unit. Several of these units may be placed in one plant to attain the electric power output needed.

Portable Power

Sometimes engines are small enough to be transported relatively easily, and may be mounted into large trailers and used for emergency power wherever they are needed. They can then be driven to a site and hooked up to generate power in a matter of days. (Fig. 4-1)

For example, during the summer Special Olympics in Raleigh, NC, the Cat Rental Power dealer network was called upon to provide power and temperature control equipment for the more than 450,000 athletes, coaches, family members, volunteers and sponsors. The local Cat Rental Power dealer in North Carolina supplied more than 8.5 MW of power and more than 1,250 tons of temperature cooling during the games. Competitions were held at venues throughout Raleigh, Durham, Chapel Hill, and Cary. (Fig. 4-2)

The local dealership powered lighting, sound, and television equipment during the opening ceremonies with one generator set package, rated to provide 300 kW of continuous power, another 350 kW genset, and three 1,250 kW gensets.

Fig. 4-1 *Cummins Power offers a variety of pre-packaged diesel-powered rental units.*

Fig. 4-2 *Portable diesel power generators cool the gymnasium during the summer 1999 Special Olympics.*

During the games, three gensets, each rated to provide 70 kW, and two gensets, each rated to provide 225 kW, powered food booths, vendor display areas, misting tents, medical trailers, logistics trailers, hospitality tents, and a Jumbotron video screen. In addition, the local Cat dealer provided one 350 kW twin pack and three 85 kW generators to power the closing ceremonies at Wallace Wade Stadium. (Fig. 4-3)

Air conditioning was at a premium during the events. To keep the athletes and spectators comfortable, Cat supplied 1,250 tons of air conditioning via chillers, air handlers, and direct expansion cooling system units.

Barge Mounted

Another popular use for small engines is barge-mounted power. Several diesel engine-and-generator sets can be mounted to a large barge and floated to countries such as the Philippines or Indonesia where they are used as distributed generation facilities. This particular application is not yet used in

Fig. 4-3 *Caterpillar rental units installed outside Wallace Wade Stadium to provide electricity and cooling for the week-long Special Olympics.*

the United States, although it certainly could be in the future, especially in cities with ports or rivers.

Such facilities provide electricity very quickly. They can be made ready to produce electricity in a matter of weeks or months, rather than the years it takes to permit and build a utility-scale power plant. Some are used to provide power temporarily due to maintenance or drought while others are considered permanent installations.

In some areas of the world, engines are popular for power generation because they can be modified to burn a variety of fuels. This can be a great asset in a country with little infrastructure or where it can be difficult to obtain fuels.

Peaking Power

Diesels are good candidates for utility's peaking power needs. Rather than building a large, capital-intensive gas turbine facility, some utilities choose to cover that load with reliable diesels.

When Sussex Rural Electric Cooperative's largest commercial customer needed standby power and ways to reduce its power expense, the New Jersey-based utility found that diesels could meet this important customer's needs while also assisting the utility's peak-shaving efforts. Sussex installed a 1,250 kW Cummins Onan PowerCommand generator. (Fig. 4-4)

Power interruptions can be a major problem affecting production lines for the Ames Rubble Corp., a supplier of rubber parts to international high-tech, business machine, photographic, and automotive manufacturers. The company's electric service peaks are about 1 MW, which can be as much as 5% to 7% of Sussex's entire load. The PowerCommand generator was the option of choice for several reasons.

"Price, reliability, and reputation were some of the main reasons we chose Cummins Onan. Another big reason was the opportunity to remotely monitor and control the genset from our office using PowerCommand software for Windows," said Jim Siglin of Sussex. The utility monitors peaks and can run the generator at the customer's site accordingly. A personal computer at the utility is linked via phone line to the off-site generator. It receives

Fig. 4-4 *A 2 MW Cummins Onan PowerCommand Genset.*

notification of any "off-normal conditions" – indication that the generator is running, that it is low on fuel or notice of a breaker operation, for example. If the alarm is not acknowledged by someone at the utility after approximately one minute, the control PC sends an alphanumeric message to individual pagers carried by technical service supervisors at the utility.

"The beauty of the software is that once you're connected to the generator, the software is designed so that you're looking at the generator's panel. All of the displays are recreated – it's as if you're standing in front of the unit without the noise," said Chris Reese of Sussex.

Since its installation, the generator has been used for standby power when a utility transmission line was out of service. The system worked successfully, allowing Ames to keep operation running for several hours while the transmission line was repaired.

Both the utility and its customer enjoy the savings from the generator. Ames' use of the generator allows it to cut power costs by up to 25%. In turn, Sussex has reduced its own power supply costs by about 5%. Sussex's ability to monitor monthly peaks enables it to peak-shave with its own power supplier, reducing coincident peak costs.

The economics and service benefits are key selling points for Sussex to promote its flexibility in designing custom plans for members of the cooperative.

Chicago's Commonwealth Edison (ComEd) has also turned to diesels for peaking needs. The summer of 1998 brought an acute power shortage that adversely affected the upper Midwest. ComEd contracted Aggreko to provide 20 MW of temporary diesel generation to supplement the utility's capacity. That project was deemed such a success that Aggreko was contracted to supply 60 MW of temporary diesel power the following summer.

Two Chicago substations were supplied with 1,250 kW generators totaling 30 MW each for the summer of 1999. The sites were manned by Aggreko technicians for the duration of the project.

When needed, each generator synchronizes with the grade and is base loaded to 1,000 kW. The Aggreko GreenPower generators ran six to eight hours daily, based on area grid demands.

The generators were set up in pairs and connected to one 2,500 kWA transformer. A 2,300-gallon EnviroTank fuel tank allows approximately 15 hours of continuous run time if needed. More than 24,000 feel of low voltage cable was used to set up the peaking diesels which ran as a distributed generation peaking power solution from April to September.

Utility Expansion

Diesel-powered gensets are a modular option for utilities that need to add capacity, but not enough to merit the construction of a large facility. Since they have a small footprint, these units can often be added to existing generation sites for a capacity boost. For example, when the island of St. Kitts needed a turnkey expansion to its Needsmust power station, it selected two Mirrlees Blackstone base load gensets – an 8 MW and a 6 MW, boosting the facilities capacity by 14 MW. (Fig. 4-5)

Mirrlees Blackstone was also selected to expand the Corito Power Station on Anquilla, supplying a 16-cylinder, 2.5 MW genset. The diesels are seen as the latest technology for continuous operation in remote locations. Both of the installations discussed here included supervisory control and data acquisition (SCADA) software systems and inputs to monitor the complete generation process.

Fig. 4-5 The Mirrlees Blackstone 8 MW engine bound for St. Kitts to be installed at the Needsmust power station for base load operation.

Emergency Needs

Extreme weather events damage transmission and distribution systems regularly. No amount of preparation or maintenance can avoid these types of outages. Ice storms drop power lines through their weight. Tornadoes can be sighted from afar at night by the green bursts of light from destroyed transformers. Hurricanes blow boats up onto dry land, and power lines and towers down to the ground. When the damage is extensive, or when it is in areas where power supply is critical to relief efforts, portable power may be called into action.

Cat Rental Power helped North Carolina avert major disruptions when Hurricane Floyd hit, dumping 20 inches of rain on the state when it was already soggy from a recent bout with Hurricane Dennis. Caterpillar dealers in and around North Carolina got busy supplying generator sets and auxiliary equipment for the cleanup effort. In total, 85 MW of power and more than 150 generator sets were placed by Cat Rental Power at hospitals, schools, wastewater treatment facilities, industrial plants, food processing and distribution facilities, radio stations, and local municipalities.

On-site at Pitt County Memorial Hospital in Greenville, NC, Cat provided a sound attenuated XQ1750 power module, rated to provide 1,750 kW of power. The unit provided power for normal operating procedures at the facility, including electrical systems, air conditioning, backup and prime electrical power. Without this standby power, the facility would have faced severe complications for its patients. (Fig. 4-6)

At Centura Bank, in Rocky Mount, NC, one genset supplied 225 kW of power to keep the facility up and running at normal operating speed.

Contacted during the storm, Cat provided the necessary power and auxiliary equipment within a few hours, including delivery, setup and testing. Cat Rental Power has been servicing customers for emergency power needs for more than 10 years. The Caterpillar line of diesel- and gas-fueled engines ranging in size from 8 to 16,200 kW.

Fig. 4-6 *Pitt County Memorial Hospital counted on Caterpillar for the power to care for its patients.*

Case Study: New Ulm Chooses Diesel Peakers

New Ulm Public Utilities did not own enough generation for its projected peak period last summer. The utility was expected to need approximately 115 percent of its installed capacity to meet peak usage, so management investigated options for providing the additional power. The utility, located 75 miles southwest of Minneapolis, provides electricity to the city of New Ulm and surrounding areas, serving approximately 6,000 customers.

Faced with the choice of purchasing additional capacity from other utilities or agencies to provide the needed 15 percent cushion, the utility turned to Minneapolis-based Ziegler Power Systems for a 4 MW Caterpillar Rental Power system.

According to Daniel Sonnek, P.E., New Ulm planning and development engineer, the decision came down to cost, flexibility, and accessibility. The proximity and prior service provided by Ziegler, along with access to the size of equipment needed, were a few of the reasons cited for choosing Ziegler.

"The monthly cost of leasing equipment was greater than normal capacity contracts with an independent power provider. However, these contracts required that capacity be purchased in six-month blocks," Sonnek said. "For New Ulm, the majority of its peaking takes place in a three-month timeframe. The flexibility of leasing equipment made it a more attractive option."

Fig. 4-7 *New Ulm's substation with rental units installed.*

In addition, the utility avoided paying transmission fees normally associated with generation because the rental power was connected to the utility's system. With on-site generation, the utility controlled it directly, for greater accessibility. "With other arrangements, transmission curtailments and equipment failures that are not under utility control can easily prevent the contracted capacity from reaching New Ulm," Sonnek said. "The biggest challenge was getting the utility to view capacity with a different perspective. Once that was accomplished, this project proved to be a very good fit – both financially and technically." Sonnek estimates his utility saved 15 to 20 percent over a comparable capacity contract.

The configuration included two Cat 3516 power modules, rated to provide 1,825 kW of prime power, a 5,000 kVA step-up transformer and a containerized switchgear trailer. The two-year contract called for supplemental capacity during July, August, and September of 1999 and 2000. Setup and troubleshooting took approximately one week as the units were installed at a utility substation. (Fig. 4-7)

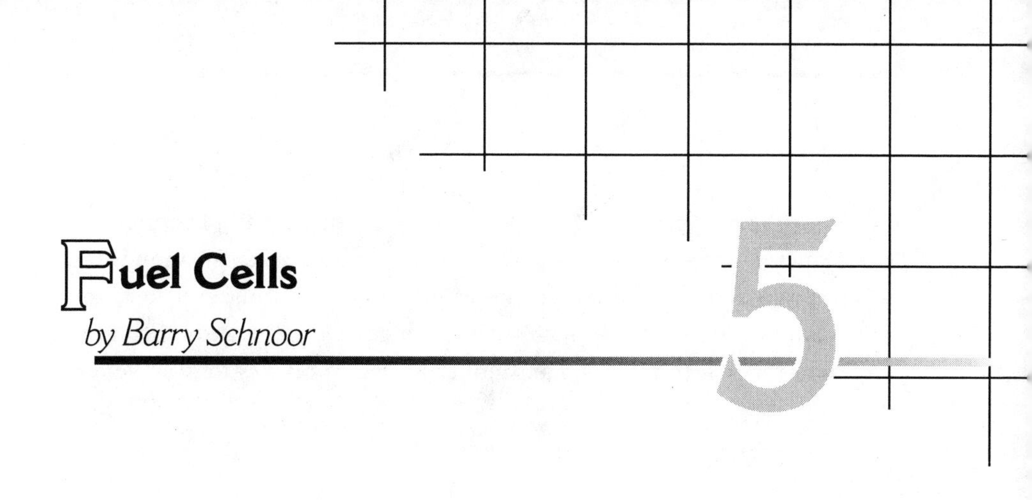

Fuel Cells

by Barry Schnoor

Global demand for energy will undoubtedly increase as we begin the third millennium. Estimates of growth rates range from 40% to 200% over the next 10 years. As the demand for energy grows, so grows the pressure on our environment, our energy resources, and our quality of living.

Meeting the energy needs of our expanding global economy and population will be a challenge in the coming decades. The intangible costs of our energy production and consumption will become more obvious. Even today, our attempts to meet basic human needs in terms of clean air are failing in cities such as Sao Paulo, Mexico City, Houston, and Los Angeles. This pollution has been implicated in premature deaths, and billions of dollars in healthcare and lost production due to illness.

Matters of the United States' national security are also an issue with the enormous dependence the United States has on foreign fuels, especially crude oil. Estimates of the world's oil reserves range from 50 to 200 years, depending on differing assumptions of future energy consumption and future extraction technology.

Regardless of where scientists ultimately sit on the issue of global warming, they have raised irrefutable arguments to change the way we are doing business today.

With the deregulation of the power industry, there is ample opportunity for emerging technologies to pave the path of growth. Deregulation is sure to open the door to distributed power generation. During and following deregulation, we may see an increase in on-site power generation, including fuel cells, ranging from power and utility related industries to a single residential unit.

Fuel Cell Basics

Fuel Cells are direct fuel-to-energy power generators that provide high quality direct current (or alternating current, after an inverter) for base load, emergency power, end-of-the-line power support, portable power, and transportation. The fuel-to-energy conversion is unlike conventional power generators.

A conventional generator uses a fuel combustion process, followed by the conversion of thermodynamic and mechanical energy (usually in the form of some sort of heat engine) to electrical energy. Because energy conversion is never 100% efficient, power is lost in each energy conversion step.

Fuel Cells operate like a battery with an electrochemical reaction. Because there is no mechanical prime mover in a fuel cell, there is no power efficiency loss due to the friction and motion of the prime mover associated with the combustion process. And because there is no combustion process, there are no byproducts of combustion either.

Fuel cells differ from batteries in several respects, however. Batteries store energy. The amount of energy available from a battery is dependent on the amount of chemical reactant stored in the battery. Once the chemical reactants are consumed (i.e., the battery is discharged) the battery will cease to produce electrical energy.

Fuel cells do not store energy. A fuel cell will continue to provide electrical power as long as it is supplied with fuel; generally hydrogen or natural gas, though methanol, gasoline, diesel fuel, and a number of other hydrogen-based fuels are being investigated as fuels for fuel cells.

Without combustion, emissions from fuel cells are inherently cleaner than associated emissions from gas turbines, internal combustion engines, or other fossil fuel burning plants. In fact, when pure hydrogen is used as a fuel, the only effluents are power, heat, and water. It is for this reason that fuel cells have been used for decades in NASA's space program. The combination of oxygen and hydrogen produces power, heat, and water — all products that astronauts require.

Fuel cells can operate on gasoline, diesel oil, methane, plant matter (bio-gas), or other hydrocarbons as their fuel stock, but the only fuel the fuel cell actually uses is the hydrogen from these fuel stocks.

A typical fuel cell plant is actually made of three independent but connected parts. (Fig. 5-1) At the user end, there is typically a power conditioner or inverter designed to take voltages and currents produced within the cell stack and make them useable for the customer. These power conditioners can be tailor-made for individual customer needs.

In the middle is the fuel cell stack itself, which is no more than a collection of individual fuel cells electrically connected in series. (Fig. 5-2)

In a fuel cell plant operating with a fuel other than pure hydrogen, there is some type of fuel processor or reformer on the front end, designed to make high quality hydrogen fuel out of a fuel that is normally too "dirty" for a fuel cell.

Steam reforming is a process in which steam and methane (or another fuel) are combined at high temperature and pressure, initiating a chemical reaction creating hydrogen and carbon dioxide. The hydrogen is then used to convert energy within the fuel cell.

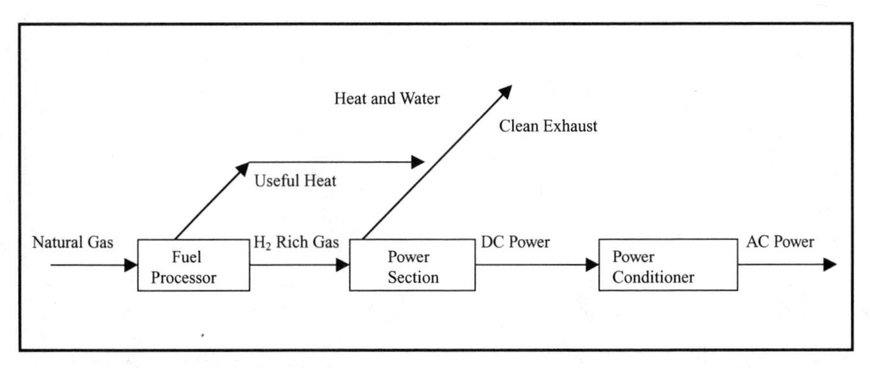

***Fig. 5-1** A typical fuel cell plant.*

Steam reforming is only one way to create the hydrogen for a fuel cell. Other methods include electrolysis and extraction of hydrogen from other industrial processes.

Fig. 5-2 MC Power's fuel cell stack.

How They Work

The theoretical basis of the fuel cell was explored by Sir William Grove before the American Civil War (1839). Having discovered that water could be decomposed into its component elements through the application of electricity (recall the simple high-school chemistry experiment in which battery terminals are connected via electrodes to a beaker of water), Grove proved it was possible to reverse the process to produce an electrical current.[1]

The basic structure of a single fuel cell consists of an electrolyte layer in contact with a porous anode and cathode on either side.[2] This single cell produces a small amount of power. Commercial fuel cells are actually a collection of "stacks" of these individual

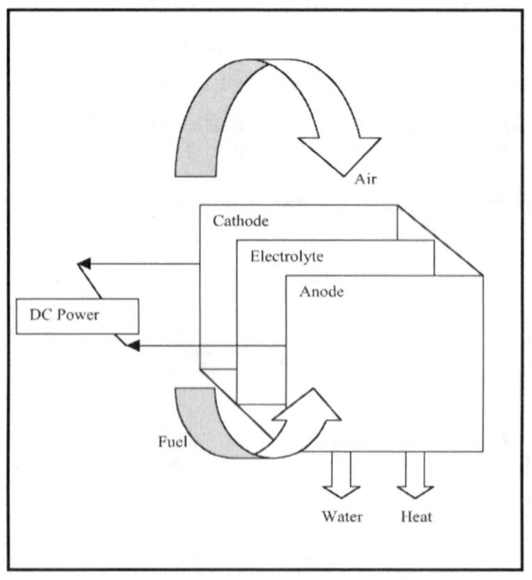

Fig. 5-3 Fuel cell stacks.

cells. (Fig. 5-3) In these stacks, cells are connected in electrical series to obtain a voltage that is practical for the given load.

Hydrogen, or a hydrogen-rich fuel, is introduced into the fuel cell at the anode of each cell. Here a reaction occurs in which the hydrogen atom splits into a proton and an electron. Freed electrons exit through the external electrical circuit as DC electricity, while hydrogen ions (the proton) pass through the electrolyte to the cell's cathode. The flow of electrons returns to the cell stack at the cathode of each cell. In the cathode, returning electrons react with hydrogen ions and oxygen from the air to form water.[3]

The electrochemical reactions occurring in phosphoric acid fuel cells are

$H_2 \rightarrow 2H^+ + 2e^-$ at the anode.
_ $O_2 + 2H^+ + 2e^- \rightarrow H_2O$ at the cathode.
The overall cell reaction is:
_ $O_2 + H_2 \rightarrow H_2O$

In theory, any substance which can be chemically oxidized can be supplied continuously (as a fluid) as the fuel at the anode of the fuel cell. The oxidant can be any fluid which can be reduced at a sufficient rate.

Hydrogen is the obvious fuel of choice for most applications, because of its high reactivity when the proper catalyst is used and because of its ability to be produced from common hydrocarbons.[4]

Ambient gaseous oxygen is, of course, the oxidant of choice for a fuel cell because it is easily and economically obtainable. Air is composed of approximately 21% O_2. Using air, rather than pure O_2, reduces the current density. Polarization at the fuel cell cathode increases with an increase in O_2 levels.

In fuel cells with liquid electrolytes, the reactant gases diffuse through an electrolyte film that "wets" portions of the porous electrode and reacts electrochemically on the electrode surface. Too much liquid electrolyte will "flood" the electrode, restricting the diffusion of gases. This reduces electrochemical performance at the porous electrode. Much effort has been devoted to reducing the thickness of fuel cell components while improving the electrode structure and the electrolyte phase. This will provide higher and more stable electrochemical performance.[5]

Fuel Cell technology is not new. In fact, it was used in a practical application during NASA's Apollo and Gemini spacecraft programs in the 1960s (just a decade before the oil crisis of the 1970s.) Fuel cells were chosen for the space program over more risky nuclear power and more expensive solar power.

Today's technology includes several different types of fuel cells, each having its unique characteristics and application.

Fuel cell types are characterized by their electrolyte, though the operation is fundamentally the same from one type to another. Phosphoric acid, molten carbonate, solid oxide, and polymer electrolyte membrane are the fuel cell designs currently in use or development for transportation and distributed power generation. Alkaline fuel cells (AFCs), used by NASA, use alkaline potassium hydroxide as the electrolyte. These types of fuel cells are currently far too expensive for commercial applications, though changes in technology may lower costs.

Polymer electrolyte membranes, or proton exchange membranes (PEMs) are typically developed for smaller applications, though Ballard Power Systems of Canada is in the production stages of a 250 kW prototype PEM fuel cell power plant.

According to the U.S. Department of Energy (DOE), "[PEMs] are the primary candidates for light-duty vehicles, for buildings, and potentially for much smaller applications such as replacements for rechargeable batteries in video cameras" and lap-top computers.[6]

These applications, particularly residential use, will have a significant impact on distributed generation. In areas where natural gas is cheap and electricity is relatively expensive, power companies may find themselves negotiating a number of net-metering or cogeneration contracts with their customers who own residential fuel cells. Plug Power, LLC predicts commercial availability of residential units by early this century, at a cost to consumers of between $3,000 and $5,000.[7]

In fact, Plug Power unveiled its first residential fuel cell during the summer of 1998 in Latham, NY. The 7 kW power plant operates solely on hydrogen, though the commercial units are expected to operate on natural gas, methane, and propane.

Phosphoric acid fuel cells (PAFC) are the only commercially available type of fuel cell as of this writing. Recent demonstration projects and year-long test operations studied later in this chapter indicate that molten carbonate (MCFC) and solid oxide (SOFC) fuel cells will be entering the market soon.

PAFCs generate electricity at more than 40% efficiency, in theory. Operational tests have indicated site performance at 37.2% (Higher Heating Value), taking into account fuel cell idle time.[8] This efficiency is not significantly higher than many of today's gas turbine generators. Like gas turbines, PAFCs can approach 85% total efficiency in cogeneration applications using heat this fuel cell produces.

Unless a fuel cell is using pure hydrogen gas, the fuel has to be reformed prior to use in the fuel cell. PAFCs require reformation external to the cell. The external reformation is required because the internal operating temperature of the fuel cell plant is not sufficient to reform the fuel internally. Carbon monoxide must be diluted to below 5% by volume to avoid poisoning the platinum catalyst.

United Technology subsidiary, ONSI Corporation, has more than 100 operating PAFCs worldwide. ONSI's PC 25 fuel cell plant is a 200 kW stand-alone package unit measuring 18 ft x 10 ft x 10 ft. Its compact size and extremely quiet operation (conversation is possible at normal volume levels immediately adjacent to the unit) allows this unit to be co-located with the load, whether that load is a communications center or a university library.

Note the footprint of this generator (180 square feet) is rather large in terms of square feet per kW. Diesel generators and gas turbines provide a much higher number of kW per square foot than does this PC 25. However, combustion generators often require sound-dampening enclosures or buildings, which increase their footprints. Even with enclosures, combustion generators are often too noisy to locate immediately adjacent to their loads.

PAFCs are not ideal candidates for emergency power. PAFC owners tend to use them for base load or primary power, tapping into the grid for emergency back-up and peak loads. The nature of the PAFC limits the start-up (heat-up) time from ambient to operating temperature. Typical start up requires three hours.

Due to the low operating temperature of PAFCs (as well as PEMs and AFCs), noble metal electrocatalysts, such as platinum, are required to produce adequate reactions at the anode and the cathode. Hydrogen is the only acceptable fuel for use within the fuel cell. Fuels from which H_2 can be derived (methane, methanol, gasoline, etc.) must be internally or externally reformed before the fuel is usable in a low temperature fuel cell. Carbon monoxide contaminates or "poisons" the anode in low temperature fuel cells, but can be oxidized in high temperature fuel cells using metals such as nickel for an electrocatalyst.

Molten Carbonate Fuel Cells (MCFCs) are expected to deliver fuel-to-electricity efficiencies of 50% to 60%, independent of plant size. Because of the high operating temperatures (about 650 C), MCFCs are more flexible in the type of fuel they can use.

The higher temperature, (compare to PAFCs at about 200 C) allows easier and less complicated fuel reformation. Reforming can take place internal to the plant, resulting in a significant efficiency increase. Carbon monoxide, a catalyst poison in PAFCs, is a directly usable fuel in MCFC plants. Cell reactions occur with nickel catalysts, rather than the more expensive precious metals. MCFCs are expected to be able to consume gasoline, diesel, and coal-based fuels such as gasified coal. This fuel requires stringent cleanup to meet the requirements of a fuel cell, however. This is a difficult and costly process.

As a result of the high internal operating temperature in an MCFC, the heat rejected from the plant is sufficient to drive a gas turbine or produce steam for use in a steam turbine for cogeneration.

Carbon dioxide is produced from fuel cells using carbon-containing fuels. While CO_2 is believed to be a contributor of global warming, note that the CO_2 produced in a fuel cell reaction is a mere fraction of pollutant-to-power ratio when compared to today's combustion processes. In MCFCs, CO_2 is used in the cathode reaction to maintain carbonate concentration in the electrolyte. Thus, CO_2 is produced in the anode, and consumed at the cathode

M-C Power and Energy Research Corporation have both developed and tested MCFCs with varying results and degrees of success. Results of these operational tests will be discussed later in this chapter.

The Solid Oxide Fuel Cell (SOFC) can be distinguished from other fuel cell designs by its high operating temperature and its solid state ceramic cell structure. The Westinghouse SOFC stack design differs from other fuel cell stacks by the absence of the high-integrity seals between cell elements, cells, and between the stack and manifold.[9] (Fig. 5-4)

SOFCs promise efficiencies of 60% in large, high-power applications such as mid-sized power generating stations and large industrial plants. (See SOFC field study later in this chapter) This type of fuel cell uses a ceramic material, rather than a liquid electrolyte. This material lends itself well to the high operating temperatures (1,000 C) and fuel flexibility expected in an SOFC.

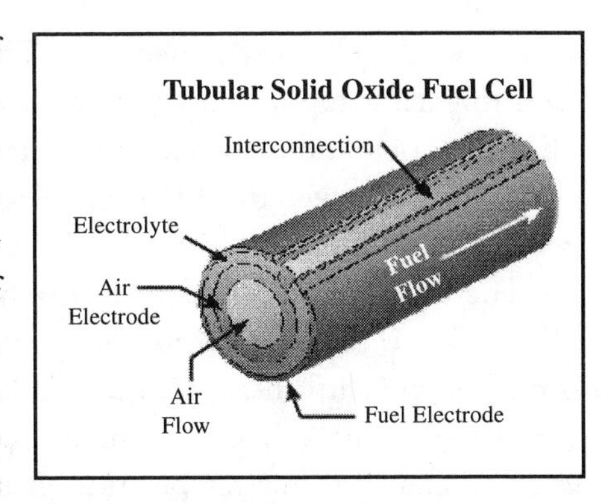

Fig. 5-4 *The Westinghouse SOFC.*

The SOFC was under development in the 1950s, even before NASA developed the alkaline fuel cell.

Similar to the MCFC, carbon monoxide is a usable fuel for an SOFC. Unlike an MCFC, however, CO_2 is not required at the cathode.

The high operating temperature allows for internal fuel reforming with the addition of reforming catalysts. Again, the high temperatures make this fuel cell plant an ideal candidate for cogeneration applications.

Future applications and renditions of the SOFC power plant will likely be linked to a gas turbine in a combined-cycle application that could achieve efficiencies as high as 75%, or even 85% when waste heat from the process is used.

The high operating temperature has some drawbacks, however. Thermal expansion mismatches among materials and seals between cells is difficult in flat-plate configurations.[10] Design changes, such as casting the cell into different shapes, may help alleviate this problem.

A 100 kW SOFC test has been ongoing in Europe and two smaller, 25 kW SOFCs are online in Japan.[11]

Polymer Electrolyte Fuel Cells (PEFC), or Proton Exchange Membranes, (PEM) function without the addition of an electrolyte, other than the membrane itself.[12] The cell consists of a proton-conducting membrane sandwiched between two platinum porous electrodes. The electrochemical reaction is similar to that of the PAFC.

Because of the characteristics of this type of cell, a low operating temperature of about 80 C is possible. The PEM is also able to sustain operation at very high current densities. This leads to a fast-start capability, ideal for many applications.[13] Because there is no liquid electrolyte in a PEM, its orientation is irrelevant to its power producing capability. This characteristic makes the PEM an ideal candidate for terrestrial vehicle power applications.

There are drawbacks of the relatively low operating temperature of the PEFC. One is the inability to use the reaction heat in cogeneration mode. Because there is little heat produced, endothermic reforming must take place external to the fuel cell. Carbon monoxide is a poison to the platinum plates in a PEFC at this low temperature. Even reformed hydrocarbons contain about 1% CO. Additional CO removal is required because the PEFC is sensitive to CO even to the low PPM level.[14]

Applications

Like many emerging technologies, fuel cells are currently too expensive for most power generating applications. ONSI's PC 25 sells for about $600,000 ($3,000/kW). Compare that with $200/kW for a typical diesel generator, and you can see fuel cells have a niche market only. Currently, their application and appeal are limited.

Fuel cells are already being used in healthcare facilities, office complexes, hotels, schools, and communications stations due to their high quality power and reliability. Heat rejected from fuel cells is successfully being used for domestic hot water, space heat, and low-pressure steam generation.

For distributed power applications, fuel cells are also being used to reclaim and make use of landfill gas. A Groton, CT, landfill is using an ONSI PC 25c fuel cell (and associated fuel reformer) to generate electricity from the landfill gas, nearly half of which is methane. While the initial start-up cost exceeded $1 million, the project supplies enough electricity to power 50 homes and the fuel is free.[15] A similar application can be made for wastewater and sewage treatment plants.

Fuel cells are also being studied and developed for personal use and transportation use in both the public (municipal buses) and private automobile realms, though fuel cell use in transportation is outside the scope of this chapter.

Stationary Generation

Some time between the realization that cities did not want smoke belching power plants near civic or residential centers, and the advent of highly efficient, low emissions power generation, most power generation was located outside city limits. Small power plants could not compete with larger, centralized power plants. It was cheaper (and often still is) to buy power from large utility plants than to produce it on-site.

Large power plants were built away from cities along major commercial corridors such as rivers and railroads. As a result, the concept of district energy went to the wayside. Process heat became wasted heat, wasted money, and increased emissions.

Decentralized or on-site power generators will become a more common choice in the future. Utility restructuring will offer utilities the opportunity to meet customer demands without the large capital expenditures normally associated with fixed infrastructure.

A recent study by Kline & Co. Inc., on the future of stationary fuel cells places the demand for fuel cells between 2,500 and 6,000 MW by 2010.[16]

Because fuel cells are extremely quiet and offer near-zero emissions, they can be located in residential areas, business parks, universities, and healthcare complexes, affording the opportunity to make use of heat that would be wasted (exhausted into the atmosphere or cooling ponds) in conventional power plants. Recent command and control technologies have made practical the use of smaller distributed power generating plants.

This heat can be used for building space heat, creating steam for cogeneration, domestic hot water, or absorption chillers. Because these generators are co-located with their loads, capturing and using this "waste" heat is simple and inexpensive.

Stationary fuel cell plants ranging from residential to the upper kilowatt capacity (perhaps even into the low megawatt capacity) are those best suited for cogeneration operation, as they can easily be sited with the electrical and thermal demand.

Plants in excess of 1 MW will likely be built for dispersed power use, taking advantage of higher electrical efficiencies and disregarding thermal energy losses. At best, these large plants may use the thermal energy produced in the fuel cell to reform the fuel or gasify the coal. High temperature fuel cells (SOFCs and MCFCs) are within range of coal gasifier operating temperatures.

Efficiencies and Environmental Issues

It is easy to understand that a 3% or 4% increase in power generation efficiency will result in a commensurate decrease in fuel consumption. Whether the fuel is gas, oil, coal, or biomass, a reduction in fuel cost translates to a reduction in overhead and an increase in the energy-to-dollar ratio for industry and business.

Of course, comparing a tried-and-true coal-fired plant with a yet-to-be-commercially-tested SOFC is difficult. Even if SOFCs were commercially available today, their capital cost, in comparison to a new coal-fired power plant, would be high in terms of cost per kW capacity. Market fuel costs for natural gas (or another hydrogen-rich fuel) are currently higher than for coal.

On the other hand, the high cost of emissions controls in a 30% efficient coal plant make it difficult to *not* justify the capital purchase of a 60% efficient SOFC plant requiring no emissions controls, and whose waste heat can be used for district heating or absorption cooling.

Something that may be less tangible is the other benefit of increased efficiencies. Let us assume that emissions per unit of fuel consumed are equivalent to conventional combustion-process power generators. A 4% increase in fuel-to-electricity efficiency translates directly to a 4% decrease in emissions; and emissions are on everybody's minds these days.

However, emissions from fuel cells are significantly lower than those of combustion generators. We know, for example, that a fuel cell operating on hydrogen has no emissions other than water vapor. We also know that a fuel cell operating on natural gas (a less expensive and more-readily-available cousin of pure hydrogen) emits only a fraction of Federal New Source Performance Standards for nitrogen oxides (NO_x), carbon monoxide, and non-methane hydrocarbons.

According to one report, "total air emissions from a [ONSI] PC25 when operated at 200 kW for an entire year adds up to 69 pounds of pollutants

... The PC25, when operated at rated power, will save more than its weight in pollutants each year it operates."[17]

In fact, PAFCs, installed in several Kaiser-Permanente locations in California, have been granted a blanket exemption from emissions permitting in California's South Coast Air Quality Management District. Exhaust from one fuel cell plant in Southern California is actually measurably cleaner than ambient Los Angeles basin air. In effect, the more power produced by fuel cells there, the cleaner the air. (Table 5-1)

Fuel cells could significantly reduce ambient air pollution while reducing dependence on foreign energy sources. There is always a concern over security of employment in traditional sectors. Jobs, such as those in the coal industry, which offer a questionable quality of life for employees and their home environments, could be replaced with thousands of high-quality jobs that actually increase the quality of life and improve the cleanliness of our homes. (Note: The same job-security concerns were raised with the advent of the telephone and personal computer. In fact, the telecommunications and computer industries comprise an enormous part of today's economy.)

Energy conversion devices, which are currently considered conventional, rely on combustion to produce electrical energy. The fuel burned is usually fossil fuel. Heat engines are inherently limited in their efficiencies by the Carnot cycle.

The Carnot cycle suggests the theoretical efficiency of a heat engine increases as the heat source temperature increases, and the temperature of the heat sink decreases (increased ΔT). In practical applications, however, the efficiency of Carnot cycle heat engines is typically less than 40% and often much less.

Emission Component	All U.S. utility generation average	Internal Combustion Generator Set (200/kW)	ONSI PC25 PAFC Factory Test
NO_x (1lb/MWh)	5.23	<3	0.016
CO (1lb/MWh)	.23	<3	0.023
SO_x (1lb/MWh)	11.01	n/a	0
CO_2 (1lb/MWh)	2,390	1,250	1,130

Table 5-1 *Emissions Comparisons.*

Fuel cells are not limited by the characteristics of the Carnot cycle. Therefore, a higher percentage of chemical energy in the fuel is converted into electricity. Heat, another by-product, is useful in other applications such as cogeneration.

The fuel-to-electricity efficiency of 40-60% (natural gas lower heating value) may be the most attractive feature of fuel cells. Fuel cells operate at near constant efficiency, independent of size or capacity. The fuel processor efficiency is size-dependent, however. Small fuel cell power plants using externally reformed hydrocarbon fuels have a lower overall system efficiency due to the parasitic electrical load of the reformer.[18] This is not a factor in fuel cells using pure hydrogen.

Because fuel cells are actually a collection of individual cells stacked one on another, fuel cell power plants can be designed to meet a wide range of electrical loads; from video cameras to automobiles to high-power electrical generation.

DOE Support

The U.S. Department of Energy budget request for Fiscal Year 2001 (FY01) programs contains well over $100 million dollars of funding for fuel cell related programs, including the following:

- **Stationary Fuel Cells** — *$42.2 million* —Programs to be funded in this section will continue R&D to reduce costs and improve performance leading to commercial fuel cell power systems within three years. In FY01, the program will begin testing a 300 kW to 1 MW size prototype solid oxide fuel cell distributed power generator at a commercial site. Funding will also focus on developing hybrid fuel cell power systems and reducing fabrication costs. The request is 12.2 percent higher than this year's enacted funding for this program.
- **Transportation Fuel Cells** —*$41.5 million* —Through the Partnership for a New Generation of Vehicles (PNGV), the FY01 budget includes funding for inte-

grating fuel cell stacks with fuel processors and balance-of-plant technologies for testing. The program will also address technology barriers to fuel-flexible fuel cell systems for automotive applications. The request is 12.1 percent higher than current funding.

- **Cogeneration/Fuel Cells** — *$5.5 million* — Formerly the Fuel Cells for Buildings program, the Cogeneration/Fuel Cells funding will go toward development of a prototype fuel processor for a low-temperature fuel cell system; completing the design competition for a 50 kW cogenerator for buildings; and continuing R&D of a membrane to operate in the 120-140 C temperature range. The request is a 55 percent jump from this year.
- **Hydrogen Research & Development**— *$23.0 million* — Funded within the DOE's Solar and Renewable Resources Technologies budget, the Hydrogen Program performs R&D of integrated hydrogen systems for power generation and transportation applications. The budget includes funding for R&D of hydrogen generation processes, technologies for fueling hydrogen vehicles, and development of a regenerative wind/fuel cell system.

The Hydrogen Program's request is 6.5 percent lower than current funding, and 18 percent lower than DOE's budget request for the program in FY2000. Bernadette Geyer, Director of Fuel Cells 2000, says "This leaves unfunded important work in cost reduction, durability testing, and validation of hydrogen generation, storage and transmission technologies." At the same time, funding requests for other advanced technology programs were far higher than current funding: Wind Energy Systems, $50.5 million, up 55.5 percent; Biomass/Biofuels Energy Systems, $102.4 million, up 44.8 percent; and Photovoltaic Energy Systems $82.0 million, up 24.4 percent.

Additional FY2001 budget programs of interest to the fuel cell community, include the following:

- **Promoting Clean Energy at Home and Abroad** — *$289 million* — This program's goal is development of technologies that convert crops and other biomass into clean fuels and other products. More than $200 million of the funding is to promote the export of clean energy technologies to developing nations.
- **Moving New Technology into the Marketplace** — *$9 billion over 10 years* — The program would provide tax relief to encourage the purchase of energy-efficient cars, homes and appliances, and the production of power from wind, solar, and biomass.
- **Advancing Clean Energy Research** — *$1.4 billion* — This would fund development and deployment of renewable energy and energy efficiency technologies for buildings, transportation, industry and utility sectors, as well as to increase research on coal and natural gas efficiency and carbon sequestration.

Technical Applications Issues

In late 1998, a task force comprised of electrical and electronic engineers, large electric utilities, and dozens of owners of distributed generation systems convened to set the standards for interconnection of distributed generation systems with the grid. Part of their goal is to adopt and adapt IEEE and UL standards for safety. One other key aspect of their ultimate goal is to eliminate anti-competitive standards designed to keep distributed generation systems (including fuel cells) from connecting to the power grid.

Interconnection With the Grid

Distributed power generation has been around for many years through private power producers and electric cooperatives. Distributed generation is now becoming increasingly popular for future power needs in the United States, in part due to the continued movement toward deregulation of the electric utility business. Perhaps the whole idea of distributed generation comes from our concept of the "American way." Self-reliance, freedom of choice, and the entrepreneurial spirit may drive the market for distributed generation. Certainly, reliable power generation and simple economics will be important factors in making the decision to move toward distributed generation.

Large power plants are capable of generating a large amount of electricity at a reasonable price, but they are restricted in terms of operating at reduced plant load. Smaller distributed generation plants, such as those discussed in this book, avoid this problem and are able to meet load demands when and where they appear. These "demand hubs" will provide peak load power, while utilities as we know them today will provide base load power. This is how many electric cooperatives operate today.

"The mini-merchant concept refers to a distributed generation facility that seeks to match its generating portfolio to a local or regional electricity demand profile in the most efficient and economic manner."[19]

Plants co-located with the demand are typically cogeneration plants. As such, they are capable of taking advantage of the overall thermal efficiency (often as high as 85%) of the plants. Because electricity and heat are used at the source, the need for separate electrical and heat plants is eliminated. Fuel consumption and emissions are reduced significantly. Compare this to the 30-35% electrical efficiency of a coal-fired power plant, with no chance for heat recovery, and it is easy to understand the benefits of distributed cogeneration.

It is important when considering distributed generation to match the type of generating equipment with its expected load. For example, a 250 kW diesel generator is ideal for peaking and emergency use, but is probably not the best choice for base loading because of its low electrical efficiency and high operating costs.

Fuel cells lend themselves to a wide variety of applications because their efficiency remains fairly constant throughout their power curve. A 200 kW PEM fuel cell may operate at 45% electrical efficiency regardless of its load. Of course, with the current high capital costs of such electrical generators, it is important to maximize the capacity of the generator. A 200 kW fuel cell may not be your best choice if your base load is 100 kW, peaking at 195 kW for one hour every day.

AC/DC

Thomas Edison was a proponent of dc applications. He favored dc because most electronic devices of his day used direct current electricity. It made sense to design a direct current electrical supply, since that is how the power is used.[20]

dc, unfortunately, experienced high line losses when transmitted over great distances. George Westinghouse discovered a way to step-up alternating current to extremely high voltages in order to transmit the power over long distances with very little line loss. It is Westinghouse's design that prompted the construction of our existing enormous power generating facilities.

Today, even modern electronic devices operate on direct current electricity and require a converter (often an inefficient, expensive converter) to change the alternating current coming out of the wall into something useful for an appliance. This is the ac power supply that is required on many modern appliances and household electronic devices, including, but not limited to televisions, microwave ovens, ranges, computers and fax machines. Fluorescent lighting and telephones also operate on direct current.

The ac power supply changes alternating current to direct current for practical use. As in any energy conversion process, this ac to dc conversion is not 100% efficient. In fact, some ac power supplies consume 20 to 40% of the total appliance load. For example, a portable television that can run off of ac or dc power may consume 70 W when plugged into a typical 110 V wall outlet, but may only consume 30 W when plugged into an automobile cigarette lighter. The differential 40 watts is consumed, or wasted, by the ac power supply, and is liberated in the form of heat.

More expensive power supplies will result in better power conversion efficiencies, but will also increase the price of the products. Manufacturers have an incentive to keep the price of their product at a minimum. Often, because the manufacturers do not pay for the operating costs of the product, they are not concerned with the operational efficiency.

It is certainly possible that 25% of all electricity consumed in residential units is actually wasted in these ac power converters.

Many renewable power producers we see today produce dc power. Photovoltaic (PV) panels and fuel cells must be connected to dc-to-ac power inverters in order to produce electricity for use in our homes. The interesting part of this power requirement is that the power then needs to be re-converted into dc power through the ac power supply. Each time the power is converted, there is an efficiency loss.

It is here that Thomas Edison's vision of distributed generation fits in. With increased use in fuel cells or PV panels, we will see an increase in distributed generation. With this, we will likely see an increase in efforts by large power producers to install barriers to distributed generation. Power generation may move from the hands of large power producers hundreds of miles away to the community fuel cell or your own rooftop.

There are design issues to be addressed in terms of the electricity consuming products in our homes, but those may have simple solutions. Eliminating the ac power supply in an appliance will increase its efficiency and may actually decrease its initial cost to the consumer. Lower initial cost. Lower operating cost. We can't lose.

Reliability

In June, 1998, we saw a combination of hot and extreme weather result in power plant outages in the Midwest, and subsequent lawsuits, price gouging and customer down-time. A surge in power demand in Western states the previous year resulted in a blackout covering several Western states.

While power reliability was intermittent in the Midwest during the summer of 1998, fatalities from the lack of air conditioning made the headlines of regional and national news. Power generators and marketers have been sued for breach of contract. We saw wholesale prices for electricity hit an exorbitant high of $7,500 MWh, nearly a hundred times the normal retail cost.

One argument against electrical power deregulation is that of stranded costs and system design. Power companies have designed their power transmission system to meet the capacity of their customers, not to serve as open-access carriers of other generators' power.

With deregulation, transmission systems will be under the common control of competing utilities with their own bottom line in mind. Until independent regional transmission companies assure access to transmission routes during times of peak demand or supply, we may continue to see extreme market reactions in extreme times.

Piecemeal distributed generation during peak demand periods will result in piecemeal power availability. Similarly, widespread distributed generation will provide widespread power availability. On-site, or regional distributed power generation has the opportunity to prevent power disruptions which result in human losses, capital equipment damage or failure, and business interruptions.

Before customers will be offered the choice of power companies, the brand of power, power companies must demonstrate they meet the prerequisites of safety, reliability, and a reasonable level of competition.

Owners of distributed generation systems will likely be expected to meet the same requirements before their generators are connected to the grid, regardless of generation capacity. In other words, a residential fuel cell unit with 8 kW peak capacity will be required to demonstrate a level of safety and reliability similar to the owner of the 1 MW SOFC and the 500 MW gas turbine plant, which also sell to the grid.

It bears repeating that a fuel cell unit may be the primary power producer for a residence or business without being connected to the grid. In this case, the grid acts as a back-up power supply and the generator does not feed the grid.

One advantage of on-site power generation in areas where grid disturbances are frequent is the high level of reliability in the power system. Operations which require high-quality reliable power will find their security and stability in on-site generation.

Safety

One of the biggest concerns of distributed generation, right down to a low voltage residential fuel cell, is safety. When power generation equipment is connected to the grid, it becomes part of the system.

Technology exists to disconnect emergency generators from the grid during power outages, so the generators feed only their primary loads, not the grid. Similar technology can be used to disconnect distributed generation units from the grid during power outages.

The concern exists for the safety of utility workers. When a power line is reported dead, workers must be assured that a small generator will not back feed into the grid during maintenance or repair operations.

Peak Shaving

If demand-side management by peak shaving (reducing peak demand, thus reducing utility demand charges) is your goal, fuel cells are likely not the best candidates. Because of the initial capital cost of fuel cells, and the nature of operation of larger fuel cell plants, smaller, more affordable generators (diesels or microturbines) are more appropriate for this type of application.

Kaiser Permanente uses ONSI fuel cells in several of its California locations. The 200kW fuel cells are used for base load, relying on the grid for peak loads. By designing the systems this way, Kaiser Permanente maximizes the use of on-site power generation without the concern or responsibility associated with selling power back to the grid. All the power and waste heat is used on site. They are effectively reducing their peak power demands and associated utility demand charges by 200 kW.

Plants that produce their own hot water and steam for process heat, but purchase electricity from the grid stand to gain significantly from on-site power generation, depending, of course, on the cost of purchased electricity.

Such a plant, which operates its own boilers for process heat, could install a heat recovery steam generator (HRSG), produce its own electricity, make use of waste heat, and exchange maintenance on the boiler for maintenance on the generator. Mid-sized 1 or 2 MW fuel cell power plants may be an ideal application for on-site power generation with heat recovery.

Thermal Considerations

Until renewable and high technology distributed energy sources are proven on the open market to meet requirements for power quality, reliability and competitive price, the minority who choose to subscribe to green power because it is the right thing to do will remain the minority.

Not long ago, gas turbine technology was believed to be best used for peak load shaving. Market and technology changes have changed that paradigm. Gas turbines, and other technologies such as fuel cells, are now being recognized as perfect candidates for base loading, particularly if an energy user can produce base load power at a price at or below the price of purchased power.

"If we can judge by the big investments major companies like Allied Signal, General Electric, Westinghouse, United Technologies and others are making in small gas turbines, fuel cells, etc., there are a lot of knowledgeable people who are betting the old rules are crumbling."[22]

Because distributed generation, including some types of fuel cells, offers a simple and inexpensive solution to the question, "How can we make use of this waste heat?", it also offers an exciting opportunity for a reduction in overall plant and facility energy use.

HRSGs can produce the base load required for a building, while the heat which used to be referred to as "waste" can be used as process heat or for building or hotel services. This use of waste heat reduces the need to burn additional fuel for process heat, thereby reducing plant emissions and operating costs. This is where the economic and environmental benefits are realized.

In fact, thermal load, rather than electrical load, is being considered for building utility designs now. Distributed generation design, in these cases, is dictated by the thermal load, not the electrical load. Therefore, a higher percentage of "waste" heat is used on site. In cases where 100% of electrical power must come from on-site generation, consideration should be given to nearby thermal loads that may be able to make use of excess heat.

Santa Clara Demonstration Project

Fuel Cell Energy and Energy Research Corporation entered into a joint private sector/government program to operate a 2 MW MCFC at a Santa Clara, California, municipal electric site in April 1996. The plant was actually an array of 16 MCFC stacks (internal reforming) of 125 kW capacity each. The primary purpose of the project was to demonstrate the MCFC, or direct fuel cell, and to promote its viability as a commercial product. (Fig. 5-5.)

The plant, the largest MCFC test to date, achieved an electrical efficiency of 44% (lower heating value). Supplemental fuel was used to ensure stability, as this was the first test of a large MCFC plant. An estimated electrical efficiency of 49% would have been reached without the supplemental fuel.[23]

The plant reached a number of operating goals, including peak power output of 1.93 MW AC, voltage harmonic quality, 2 ppm NO_x, undetectable SO_x, and noise levels of less than 60 dBA at 100 feet. However, this successful test was not without problems.

Fig. 5-5 *The Santa Clara fuel cell installation.*

An adhesive used to attach dielectric insulators and thermal insulation to stacks and process lines deteriorated during periods of high temperature operation. The adhesive became an electrical conductor, causing damage to dielectrics and plant piping.

After some reconfiguration and repairs, the plant was brought back on line at 500 kW using the 8 stacks that had not been damaged. The reduced load on the remaining 8 stacks (they were collectively rated at 1 MW) allowed for lower operating temperatures, reducing the adverse impact of extreme heat on the stack.

The plant reached a total operation of 4,900 hours of hot time (plant at normal operating temperature) and 3,400 hours of grid connected operation.[24]

Mercedes Tries Fuel Cells

Southern Co., Alabama Municipal Electric Authority, FuelCell Energy and Mercedes-Benz U.S. International Inc. have joined together in an historic partnership to build and install a fuel cell power plant at the Mercedes-Benz production facility in Tuscaloosa, AL. The $2 million, 250 kW demonstration project is expected to enter operation in 2001.

The natural gas power plant uses FuelCell Energy's Direct FuelCell stack in a special design developed by MTU Friedrichshafen. It is expected to be

50 percent efficient. "Southern Co. is dedicated to exploring the full array of energy opportunities, including customer-sited, fuel cell power generation," said Dr. Charles H. Goodman, Southern Co. vice president for research and environmental affairs.

Southern Co., AMEA and FuelCell Energy will provide funding for the power plant, and Southern Co. is serving as project manager. The plant will operate for at least one year, and companies may agree to continue the project after the first year. The plant will feed the Mercedes-Benz facility's power distribution system. The plant will be skid-mounted for ease of transport in future demonstrations. MTUY is conducting a 250 kW commercial field trial at Bielefeld, Germany, providing steam for a nearby university, heat for district heating, and electricity for the grid.

MC-Power's Demonstration Project

Early in 1997, M-C Power Corporation completed an operational test of a 250 kW MCFC power plant at Naval Air Station, Miramar, California, near San Diego. This test was conducted as a cost sharing agreement with the DOE.

The primary goal of this Project Development Test, as it was called, was to conduct a field study of a completely integrated MCFC power plant in order to identify improvement opportunities in design, performance and cost. M-C Power had conducted a previous short-term operational test and had integrated design changes into the plant used during the Miramar test.

Despite all the problems that can be expected (though not predicted) with high-tech power plant testing, M-C Power called the test operation a success. During the test period, a number of accomplishments were achieved:[25] They include the following:

- Uniform stack voltage. The average voltage was 720 mV per cell (a total of 250 cells). Standard deviation was about 20 mV/cell.
- No electrolyte pumping. MCFC developers have experienced electrolyte (molten carbonate) migration within and across cells, resulting in power degradation. This was not a problem, even in this commercial size stack.

- Good gas seal integrity. Cathode and anode gases flow in adjacent paths on opposite sides of the separator plates in the stack. These gas seals keep the cathode and anode gases from mixing or from leaking from the stack.
- Reliable pressurized operation. One unforeseen advantage of the initial start-up problems was the ability to test the transition between stand-by and online operation numerous times. The Miramar test was a good test of the responsiveness of this process.
- Excellent reformer operation. The fuel reformer lived up to its expectations in terms of output and conversion efficiency. Operation for the five-month test was virtually trouble free.

The project test began on January 10, 1997. Two weeks later, the plant began to produce power. The anticipated problems showed up immediately. Problems with the inverter, turbocharger, and recycle gas blower caused a number of outages during the first few weeks of the operation. Future tests will include a more robust version of the turbocharger and recycle gas blower.

Inverter failures were caused by a loose set of AC power connections internal to the inverter cabinet. Power spikes from the loose connections interfered with inverter control logic. Troubleshooting identified the problem and the inverter performed well during the last couple of months of operation.

The plant operated until May 12, 1997, having completed 2,350 hours of "hot" operation (stack temperature above 600°C). The plant produced 160 MWh of dc power (maximum output of 206 kW dc) and exported about 340 MMBtus of steam at 110 psig.

The NAS Miramar plant averaged an electrical efficiency of 44.4%. Thermal energy captured during plant operation increased cogeneration efficiency by another 10%. The planned commercial unit (500 kW capacity) is expected to achieve an electrical efficiency of 52% (lower heating value of gas), approaching 70% in cogeneration mode. Thermal recovery modifications could increase cogeneration efficiencies to nearly 85%.

M-C Power feels the successful test of this robust fuel cell will allow greater flexibility in future design, bringing the corporation closer to the ultimate goal of successful commercialization.[26]

M-C Power expects the installation of up to four commercial prototype 250 kW cells early this decade to demonstrate the flexibility and reliability of this type of fuel cell.

A 1 MW CFC plant is being planned for operation at a wastewater treatment digester plant in King County, Washington. Plant startup is tentatively scheduled for late 2001.[27]

Solid Oxide Fuel Cells

A partnership between innovative private companies and the federal government working toward the development of the first megawatt-class solid oxide fuel cell (SOFC). According to Westinghouse, this SOFC can transform the power production sector around the world.

Westinghouse claims SOFC plants can achieve electrical efficiencies of up to 75% and lower heating value when used in a combined-cycle application.[28]

Westinghouse is involved with a number of SOFC projects, including EPA's Fort Meade Maryland laboratory SOFC installation.

The original proposed installation consists of two 500 kW SureCELL fuel cell modules, supplied by Westinghouse, and a 300 kW turbine generator. The combined-cycle yield of this project will be approximately 1.3 MW of electrical power.

The National Energy Technology Laboratory reports that technological and financial delays have put this project proposal off since 1997, but there is anticipation of a project "kick-off" in 2000 to breathe new life into this exciting demonstration project.[29]

Westinghouse has been involved with a preliminary design of a flexible-fuel SOFC power plant for the United States military. In conjunction with Haldor Topsoe's logistic fuels processor (LFP), the proposed SOFC power plant will produce 3 MW of electricity (nominal capacity) using DF-2 diesel fuel or JP-8 jet fuel.[30]

The feasibility of operating an SOFC power plant in conjunction with the LFP was tested with a 27 kW-scale test near San Bernardino, CA, in 1994 and 1995. The SOFC plant operated for 766 hours on JP-8 jet fuel and 1,555 hours on DF-2 diesel with fuel supplied via the LFP. Additionally, the small SOFC plant operated for 3,261 hours on pipeline natural gas. The plant produced 188 MWh of electricity over 5,582 operating hours with no perceptible degradation in performance.[31]

To achieve the 3 MW capacity, six pressurized SOFC stacks (500 kW each, sized for military truck transport) will be required. These stacks, or submodules can be arranged in a number of configurations: A six-pack with one gas turbine generator, two 3-packs with two GTs or three twin-packs, each fitted with a gas turbine. The military seems to prefer the latter design, as it is most easily transported with military vehicles.[32]

Siemens Westinghouse Power Corporation of Orlando announced in January 2000 that its 100 kW SOFC power plant, located in the Netherlands, successfully completed one year (8,760 hours) of total operating time. Steve Veyo of Siemens said the system surpassed 9,500 operating hours with no sign of power degradation. (Fig. 5-6)

The 100 kW unit, installed at a cogeneration plant in Westervoort, the Netherlands, is the world's largest SOFC plant and has the longest run-time of any fuel cell of this type and size. The purpose of the Westervoort SOFC project is to demonstrate the durability and performance of this type of fuel cell power plant.

The power plant, designed to output 100 kW of electricity, is actually exceeding its nameplate capacity, providing 110 kW of electricity to the local power grid. Additionally, the SOFC power plant is providing hot water to the local district heating system.

Without the benefit of a combined-cycle system, this power plant has reached a unit fuel-to-electricity efficiency of 46%, well below the 60% efficiency estimated just a few years ago. It should be noted that theoretical estimates and tests of pressurized SOFC (PSOFC) plant performance at fuel pressures up to 15 atmospheres, combined with back-ended gas turbine generators, economically yield 63 percent electrical generation efficiency (net ac/lower heating value).

The claim of environmental benefits seems to ring true, however, as NO_x emissions have been held to 0.2 ppm. Sulfur oxides, carbon monoxide, and volatile hydrocarbons have all been measuring less than 1.0 ppm.[33] The demonstration, operating for EDB/Elsam (a group of Dutch and Danish utilities) will be followed by further tests of different plant capacities and market applications.

Fig. 5-6 A Siemens fuel cell plant.

ONSI's Demonstration Projects

ONSI (derived from "On-Site"), a division of United Technologies Incorporated is the closest to breaking open the market for Phosphoric Acid Fuel Cells (PAFC). In fact, as of Winter 2000, the ONSI PC25 PAFC power plant is the only commercially available fuel cell plant. ONSI has more than 3 million operating hours of experience with this type of power plant, and reports operation of up to five years without stack replacement.[34]

The PC25 fuel cell (Fig. 5-7) has been available since 1992 and, after fuel reforming, is capable of operating on a number of different fuels, including natural gas, propane, hydrogen, and methane. ONSI has this model of fuel cell operating on landfill gas in Connecticut and on anaerobic digester gas from waste water treatment plants in Massachusetts, Oregon and Japan.[35]

A Westchester County, New York, wastewater treatment plant has, on site, a 200 kW ONSI fuel cell. The gas, which had previously been flared (burned off) produce enough power to keep the treatment plant running, and emissions from combustion have been all but eliminated.

The ONSI unit has been installed high atop an office building in New York's Times Square and is in operation in dozens of U.S. Military installations through energy programs in conjunction with the Department of Defense.

The U.S. Navy Education and Training Center in Newport, RI has been operating the PC25 fuel cell near its boiler plant since February, 1995. This PAFC plant is co-located with the base's No. 7 boiler plant. The electrical output is connected to the electrical transformer feeding the boiler plant. The heat from the fuel cell power plant is used to preheat boiler make-up water, reducing the amount of fuel necessary to bring the water to operating temperature.[36]

***Fig. 5-7** An ONSI PAFC.*

The Department of Defense has released the fuel cell performance figures through December 1999. They are as follows:

Hours of Operation	35,884 Hours
Total Electric Output	5,485 MWhrs
Total Heat Recovered	7,279 MMBtu
Input Fuel	60 MMCubic Feet
Electrical Efficiency	30.2%
Thermal Efficiency	11.7%

The Department of Defense also reported an annual cost savings of $94,000 at the Newport location, including the cost of electricity and natural gas. A Defense Department fuel cell demonstration project at Edwards Air Force Base, California, (using a similar ONSI 200 kW fuel cell) produced an estimated annual savings of $96,000.[37]

These large savings, and the other benefits of fuel cells, make this an attractive application in these locations. The savings are a function of fuel (natural gas) prices, and the purchase price of electricity. These savings will not be realized in locations where natural gas is expensive and electricity is more affordable.

The United States Military Academy at West Point is a perfect example. This application estimates an annual net savings of $30,000. This represents a 20-year payback on the $600,000 purchase price of the unit, even though electrical efficiencies (31.6%) and overall efficiencies (64%) are higher than the Newport plant.[38]

The National Defense Center for Environmental Excellence, located in Johnstown, PA, has been operating the PC25 unit since August, 1997. Though the efficiency figures were not immediately available, the cost savings were, at $15,000 per year. It should be noted that the estimated use of the fuel cell thermal energy is about 19%, opening tremendous opportunities to capture and take advantage of the plant's "waste" heat.[39]

Polymer Electrolyte Fuel Cells

A number of companies are pursuing PEMs or PEFCs. This type of fuel cell will likely be the next commercially available plant. The commercially available version of the PEFC will be smaller, more affordable, and more widely distributed than its phosphoric acid cousin.

Avista Labs has been developing a PEFC designed to produce power and hot water for residential units. Ranging from 2 kW to 10 kW, this fuel cell is about the size of a residential water heater and is expected to be available early this decade. Avista Labs' unit will operate on reformed natural gas.

Plug Power, LLC has been operating a PEFC fuel cell since June 1998. Their residential size unit (7 kW) initially began operation using pure hydrogen as a fuel. By August 1999, this unit was operating on natural gas. Commercially available units are expected to operate on natural gas, as natural gas is conveniently available to millions of homes in the United States.

On Valentine's Day 2000, Plug Power announced that it has successfully run a PEM fuel cell for more than 10,000 hours. This is certainly an accomplishment. To be economically feasible for residential use, however, a residential fuel cell would likely have to be robust enough to last approximately 10 years for a simple payback. This figure depends, of course, on the prices of natural gas and purchased electricity. A homeowner may be able to save money on the per-kilowatt basis, but if the fuel cell needs to be replaced every two or three years, the simple economic payback will not be there.

Plug Power, through a joint venture with GE MicroGen, hopes to have fuel cell systems commercially available beginning in 2001. The commercially available unit will have an electrical efficiency of 40%. When excess heat is captured (primarily for hot water, but perhaps also for space heat) overall efficiencies can approach 80%.[40]

Residential fuel cells are currently expected to sell for between $7,000 and $10,000 each, depending on load capacity and design. In mass production, however, these systems are expected to retail for $3,000 to $5,000.

H. Frank Gibbard, CEO of H Power Corporation, expects the price of his company's residential unit to sell for less than $3,000 by the year 2004.[41] Depending on fuel costs in a given market, such a price tag could mean that the fuel cell would produce electricity for around 7 cents/kWh. That could be very attractive for residents of high power cost areas; even grid connected customers.

H power, in a long-term partnership with Energy Co-Opportunity (ECO), hopes to open the market of residential and small commercial fuel cells to electric cooperatives and their customers.

Northwest Power Systems built a 5 kW residential unit under contract with Bonneville Power Administration. The fuel cell, owned by BPA was successfully demonstrated in a 2,250 square-foot house in Bend, Oregon.

Recently, Northwest Power Systems developed a fuel processing technology that converts diesel fuel into high quality hydrogen gas. NPS's design produces nearly pure hydrogen with less than one part per million (ppm) of carbon monoxide and less than one ppm of carbon dioxide. These contaminants can poison a PEM fuel cell irreversibly, as described earlier in this chapter.[42] (Fig. 5-8)

Fig. 5-8 *Northwest Power System's 5 kW unit.*

Ballard is among the leaders in PEFC manufacturing, and is currently demonstrating a 10 kW gas-powered fuel cell. Ballard has been designing and manufacturing PEM's for transportation applications for years (in agreement with Ford and Daimler-Chrysler, among others) and has developed a commercially available 250 kW natural gas PEFC for stationary, distributed power applications. This unit was being field tested as of the writing of this chapter. (Fig. 5-9)

Fig. 5-9 *A Ballard 250 kW fuel cell.*

The Future of Fuel Cells

"Fuel Cells: On the Verge," a study by Business Communications Company Inc., suggests the total market for fuel cells is projected to reach $1.3 billion by 2003, an annual increase of more than 29%. Fuel cells closest to commercialization will see the strongest growth, especially PAFCs and PEFCs. The report predicts a sales increase from $41 million today to $250 million by 2003, for PAFCs.

Bear in mind this is one prediction from one company. Someone once bet the farm that the Beta video tape format was the wave of the future. This technology is changing and developing rapidly, particularly in the PEFC and PAFC realms.

PEM fuel cells, those already being demonstrated as ideal for residential or transportation use, will likely grow quickly once the technology is improved and the market opens up. Automobile companies such as Daimler-Chrysler, Ford and General Motors are betting the market will grow.

Though fuel cells in use in transportation do not directly affect distributed power generation, technological improvements coming as a result of increasing the fuel cell transportation market will certainly benefit stationary units. In fact, PEFCs designed for automobile use may be close cousins to those appropriate for residential distributed use.

In addition to technological improvements offered by pioneering PEM fuel cells in transportation, the progress in cutting back mobile pollution sources, particularly in places such as California's South Coast Air Quality District, will relieve pressure to heavily regulate stationary power producers.

As is the case with most recent emerging technologies (DVDs, CD players, personal computers, cellular telephones, VCRs and microwave ovens, to name a few), the purchase price of an operating fuel cell will likely fall as demand increases and market supply meets that demand. Ford's Model-T was too expensive for most people when it was first introduced, too. There are not yet enough fuel cells produced to take advantage of the economies of scale.

Today's commercially available PAFC costs about $3,000/kW (plus $85,000 for installation depending on application, heat recovery hardware, etc.) despite estimates five years ago suggesting today's prices would be about half of what they are.

No one expects fuel cells to be available at your local hardware store any time soon. Still, it is not outside the realm of imagination to think that residential fuel cells may soon be as much a part of a building or home as today's heat pumps, high voltage transformers, and air handlers. This type of commercialization is what fuel cell manufacturers are anticipating.

The market issues of supply and demand are not the only economic issues. The primary factor in fuel cell availability is the differential between purchased electricity rates and fuel (natural gas) rates. Thermal efficiencies, utility demand charges, and perhaps emissions credits may also play a part in the financial decision.

There are, of course, technical and engineering challenges that producers of fuel cells have tasked their scientists and engineers to solve. Case studies discussed in this chapter discuss several of these demands and lessons learned.

Today's fuel cells have lasted between 5 and 7 years of continuous operation without degradation significant enough to warrant stack replacement. If simple economic payback is our only criteria for fuel purchases (if we place dollar value much higher than energy quality and environmental issues) fuel cells will have to last *at least* past the simple payback period in order to be affordable. PAFCs have been the only long-term fuel cells in operation. Other fuel cell technologies have exceeded 10,000 hours of operation, but 10,000 hours is only just over a year of continuous operation, the type of which a fuel cell would be subjected to in many industrial and residential applications. Future fuel cells will have to be robust enough to last through the rigors of, say, 100,000 continuous operating hours.

Though fuel cells generate electricity at efficiencies much greater than combustion technologies, and with significantly reduced emissions, the fact remains that the primary fuel for today's fuel cells is still fossil fuel. True, increased efficiencies should extend the time for which fossil fuels are available to us. But there are still political, economic, and environmental issues associated with the extraction of these fossil fuels.

Jules Verne predicted more than a century ago (*The Mysterious Island)* that our society would be powered by hydrogen. He was certainly ahead of his time, but he may have hit the mark dead on. Perhaps he knew what Sir William Grove knew two scores earlier.

The DOE is continuing its sponsorship of emerging technologies designed to improve the production and storage of useable hydrogen. In addition to hydrogen extraction from fossil fuels (fuel reformation), hydrogen production using photosynthetic, biological (biomass digestion) and chemical processes are being studied.

Safe storage of hydrogen has also been a concern. Storage in carbon and metal hydride or hydride slurries, as well as more conventional pressurized gas storage, is being researched and developed.

Hydrogen earned a bad reputation as an unsafe fuel in part because of the Hindenburg incident. Certainly, hydrogen played in important part in the tragedy, but some believe now the airship would have burned anyway. Retired NASA safety expert Addison Bain points to the construction of the ship's gas bags. The bags were made of either cellulose acetate or cellulose nitrate, both flammable. Additionally, the bags were coated with aluminum flakes to reflect the sunlight for thermodynamic and aesthetic purposes. Cellulose nitrate and metal chips are ingredients in rocket fuel. Perhaps the dirigible would have burned even had it been filled with inert helium.[43]

These observations are not intended to display a cavalier attitude about the potential hazard inherent in hydrogen gas. Surely we all recognize the dangers of gasoline in both liquid and vapor form. All flammables and combustibles must be treated with a certain amount of respect. The fact that hydrogen gas dissipates so quickly may actually decrease post-accident fire hazard.

Scientists are experimenting with different designs for hydrogen storage, making storage safer and more economical.

People who, only a few years ago, invested capital in companies dedicated to the development and manufacture of fuel cells have not been disappointed. Technology and market breakthroughs have been very encouraging to companies and stockholders alike. The fuel cell industry is poised to take on the challenges and the opportunities of a deregulated power industry.

Will fuel cells be the panacea? That certainly remains to be seen. In any case, fuel cells are here and show tremendous promise for the future.

With concerns over international fuel availability and supplies, increased pressure to stop off-shore drilling, increased demand for uninterrupted high quality power for high-tech industries and increased awareness of air quality and other environmental issues, fuel cells may well have found their niche. And it may be a big one.

End Notes

1. Engleman, Ron R. Jr., October 1996, *AIArchitect Magazine*, p. 13.
2. *Fuel Cells: A Handbook*, Revision 3, Morgantown Energy Technology Center, U.S. Department of Energy. p. 1-2.
3. Cler, Gerald L., March 1996, *The ONSI PC 25C Fuel Cell Power Plant*, E Source, Inc. Product Profile p. 996-2, p. 3.
4. *Fuel Cells: A Handbook*, Revision 3, Morgantown Energy Technology Center, U.S. Department of Energy. p. 1-2.
5. *Ibid.*
6. Fuel Cells 2000's types of Fuel Cells Page, http://www.fuelcells.org/fctypes.htm, 4/29/97.
7. Fuel Cells 2000 Technology Update, July 1998, http://www.fuelcells.org.
8. http://www.dodfuelcell.com/site_performance.html, 6/17/98.
9. S. E. Veyo, *Fuel Cell Power Plant Initiative: Final Report*, Siemens-Westinghouse 1998.
10. *Fuel cells: A Handbook*, p. 1-2.

11. www.fuelcells.org/fctypes.htm.
12. Grune, H., abstract from The Fuel Cell Seminar Organizing Committee, 1992 Tucson, AZ, p. 161.
13. *Fuel cells: A Handbook*, p. 1-2.
14. Hirschenhofer, Stauffer, Engleman and Klett, *Fuel Cell Handbook*, Parsons Corporation 1998 for U.S. Department of Energy, p. 6-2.
15. Hawkins, James., ONSI Corporation, Personal conversation March, 1996.
16. Fuel Cells 2000, Fuel Cell Technology Update, July 1998, www.fuelcells.org.
17. HI Disposal Systems, "Fuel Cells for Hospitals," www.hicompanies.com/hawkins_51.html.
18. *Fuel Cells: A Handbook*, p.1-17.
19. Shelor, Mack F., *Mini-Merchants for Distributed Generation*, Power Engineering, Vol. 102, No. 8, August 1998, p. 34.
20. www.energy.com, 12/1/98.
21. Sanda, Arthur P., Coal Age magazine, *Fighting Windmills, Literally*, April, 1997.
22. Zink, John C., Managing Editor, Power Engineering Magazine, *The Rules are Changing*, PennWell Publishing, Vol. 102, No. 7, July 1998.
23. Leo, A. J., O'Shea, Thomas P., Skok, Andrew J., "Santa Clara Direct Carbonate Fuel Cell Demonstration," *Proceedings of the Fuel Cells '97 Review Meeting*, FETC, Morgantown, WV, August, 1997.
24. Hirschenhofer, et al, p. 4-29.
25. Petkus, Robert O., correspondence of 10 July 1998, and "Successful test of a 250 kW Molten Carbonate Fuel Cell Power Generator at NAS Mirarmar."
26. *Ibid.*
27. Hirschenhofer, et al, p.1-17.
28. Wirdzek, Philip, US EPA, *Solid Oxide Fuel Cell Demonstration Project*, Spring 1997, p. 3.
29. Mills, Otis, National Energy Technology Laboratory, Personal conversation, 1 March, 2000.
30. Veyo, S.E., Fuel Cell Power Plant Initiative: Final Report, 1997, p.1-1.
31. *Ibid.*

32. *Ibid*, p. 1-2, 1-3.
33. DOE Techline, January 31, 2000.
34. www.onsicorp.com, 3/2/00.
35. *Ibid.*
36. John Alfano, Navy Public Works, Phone conversation, August 1998.
37. dodfuelcell.com/newport.php3, 3/2/00.
38. dodfuelcell.com/westpoint.php3, 3/2/00.
39. dodfuelcell.com/ncdee.php3, 3/2/00.
40. www.plugpower.com/news, March 1999.
41. www.energy.com/coverstory/cv120298.asp, December, 1998.
42. www.northwestpower.com, April 11, 2000.
43. Fuel Cells 2000's FAQ web page, http://216.51.18.233/fcfaqs.html, February, 2000.

Barry Schnoor Biography

Mr. Schnoor gained experience in power generation and distribution serving as Plant Engineer in the Navy and Power Plant superintendent at the University of Virginia. He holds a B.A. in Economics and a Masters in Urban Planning. His diverse educational and experiential background offers a unique insight into issues surrounding distributed power generation.

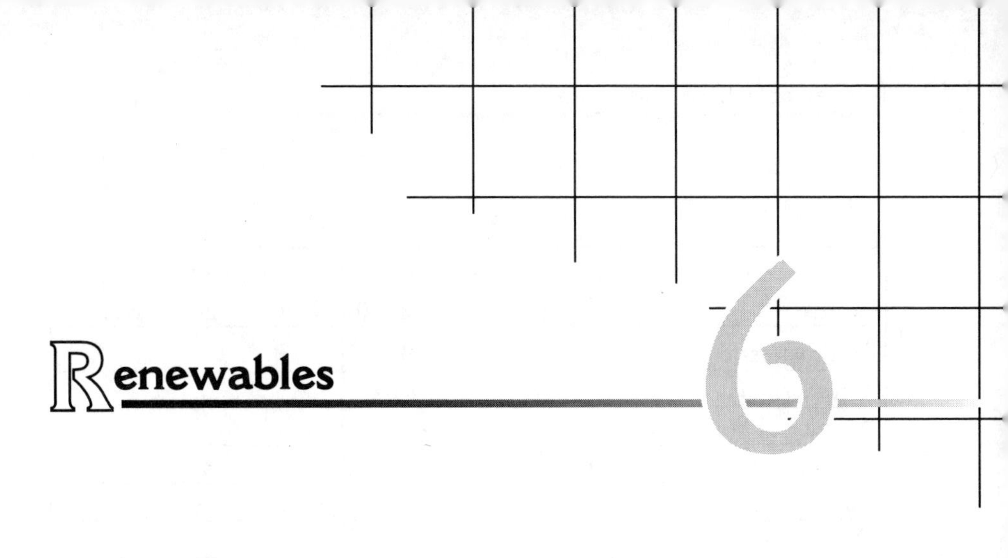

Renewables

Renewable energy sources are those that will replenish themselves – the tide, water flowing downhill in a river, the wind, and the sun. As the technologies to use these natural energy resources evolve, the expense of using renewable energy to generate electricity is coming more in line with the traditional fuels. Emissions constrictions on fossil fuels, research and development efforts, and government subsidies are bringing this niche market toward the mainstream in the United States. Government subsidies also help bring the costs more in line.

Renewables are also gaining ground in the marketing arena. They supply what is often referred to as "green power," and the U.S. public has been embracing green-power programs offered by utilities as a way to please environmentally conscious customers.

Renewable resources – solar, wind, geothermal, hydroelectric, biomass and municipal solid waste – provide about 12% of the U.S. electricity supply. Hydroelectric resources alone provide almost 10% of this, although hydro is not really a contender as a distributed generation source. Biomass and municipal solid waste (MSW) together contribute a little more than 1%. All other renewable resources – including geothermal, wind, and solar – together provide less than 1% of the total (Table 6-1, Table 6-2).

Source	Billion kWh 1990	2010	Annual % change 1990	2010, average
Hydroelectric	288	306	0.3	
Geothermal	15	62	7.2	
Municipal Solid Waste	10	54	8.5	
Biomass	31	59	3.2	
Solar	1	4	9.2	
Wind	2	16	10.4	
Renewable Total	348	501	1.8	
Fossil/Storage/Other	2,098	2,975	1.8	
Nuclear	577	636	0.5	
Total Generation	**3,023**	**4,112**	**1.5**	

Table 6-1 *U.S. Electricity Net Generation Using Renewable Resources, 1990 and 2010*

Many renewable resources are relative newcomers to the electric power market. In particular, electricity generation using geothermal, wind, solar, and MSW resources have had their greatest expansion since the 1980s. This is the result of significant technological improvement, the implementation of favorable governmental policies, and the reaction to the increasing costs of using fossil and nuclear fuels. As can be seen in Table 6-1, wind is the fastest growing renewable and resource, and likely the most viable as a distributed generation resource.

Use of renewable resources for electricity generation has also been encouraged as less environmentally damaging than fossil fuels. Because renewable energy is available domestically, renewable resources are viewed by some as more secure than imported fossil fuels.

The use of renewable resources other than hydroelectricity is increasing rapidly, particularly wind power, which has enjoyed rapidly developing technology for efficiency improvements and government incentives to bring down the expense. Conventional hydroelectric power, the mainstay of renewable resources in electric power today, is unlikely to enjoy rapid growth under current expectations, even if more favorable regulatory policies emerge.

Wind

The wind contains lots of energy. Wind turbines use the wind to create electricity. Wind power creates no pollution and has very little impact on the land. Wind energy can be produced anywhere the wind blows consistently.

Solar

The sun's radiation is used directly to produce electricity in two ways. Photovoltaic (PV) systems turn sunlight into electricity directly. Solar thermal systems use the sun's heat to heat water, creating steam to turn a turbine and generator.

Hydroelectric

Dams provide electricity by guiding water down a chute and over a turbine at high speed. Small hydropower facilities are considered renewable energy resources. Large dams are not. Hydropower does not produce any air emissions, but large dams have environmental issues such as flood control, water quality, and fish and wildlife habitat to deal with.

Geothermal

Geothermal energy is generated by converting hot water or steam from deep beneath the Earth's surface into electricity. Geothermal plants cause very little air pollution and have minimal impacts on the environment.

Biomass

Organic matter, called biomass, can be burned in an incinerator to produce energy. In some facilities, the biomass is converted into a combustible gas, allowing for from a variety of sources, including agricultural, forestry, or food-processing byproducts, as well as gas emitted from landfills.

***Table 6-2** Renewable Power Sources*

The lack of many additional large sites for hydroelectric facilities limits the potential for hydroelectric power growth.

If renewable resources are to provide a greater share of the U.S. electricity supply, costs of using them will need to decline relative to alternatives. In some cases, such as wind and solar thermal generation, small improvements in generating costs may significantly increase their market penetration. In other cases, such as photovoltaic and most forms of geothermal power, large cost reductions are needed to spur greater market penetration.

WIND

Industry estimates show more than 3,600 MW of wind generating capacity was installed in 1999, bringing total worldwide installed wind capacity to 13,400 MW. This total represents an increase of more than 36 percent over the 1998 total installed capacity of 9,751 MW and the largest worldwide addition to wind capacity in a single year, according to American Wind Energy Association records.

Between 1995 and 1998, 4,893 MW of wind capacity was installed worldwide for an average annual growth rate of 27.75 percent. Wind capacity has surged from less than 2,000 MW in 1990 to 13,400 MW at the close of 1999. (Fig. 6-1)

These statistics seem to support European Wind Energy Association claims that wind power can produce 10 percent of worldwide energy supply by 2020, even if electricity consumption increases substantially. Denmark and Germany's Schleswig-Holstein are already approaching this 10 percent figure.

"The 1990s have seen Europe emerge as a world leader in wind energy development, and we expect this strong performance to continue," said Christophe Bourillon, EWEA executive director. "Our association has set targets for Europe alone of 40,000 MW of wind capacity by the year 2010 and 100,000 MW by the year 2020."

Bourillon attributes the surge in wind power's popularity to concern about climate change, worries about fossil fuel supplies, and the need to sustain an ever-increasing population. "Wind energy can reduce the amount of greenhouse gases released into the atmosphere,

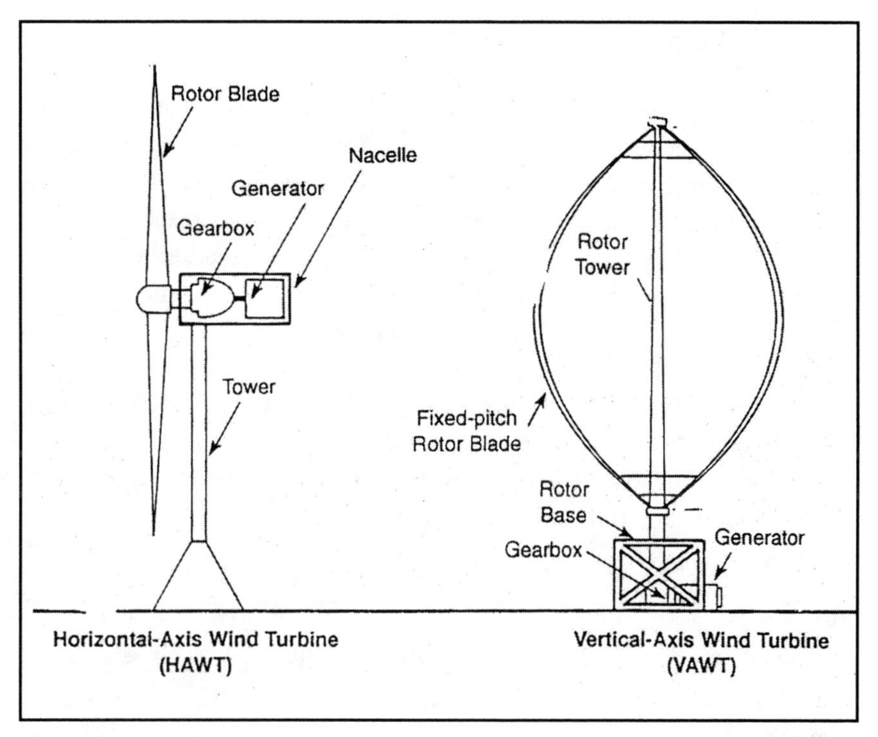

Fig. 6-1 *The two main types of wind turbines are the horizontal axis and vertical axis. Horizontal axis machines are much more common in the United States.*

preserve valuable fossil fuel reserves for specialized uses and help poorer rural countries develop without resorting to polluting technology," Bourillon said. "Although there are uncertainties because of the changeable policy environment, we are projecting more than 5,000 MW of new growth in the United States over the next decade," said Randall Swisher, AWEA executive director.

"Overall global investment in wind turbines should surpass $200 billion by 2010. The growth this past year is just the beginning of a rapid investment into renewable energy sources worldwide," said Michael Kujawa, senior Allied Business Intelligence Inc. analyst.

The top three countries added 2,582 MW of new capacity and account for almost 64 percent of total capacity additions in 1999. (Table 6-3.)

Capacity Boom

According to the AWEA's "1999 Global Wind Energy Market Report," some 732 MW of new wind capacity and an additional 173 MW of repowering projects using new turbines to replace less efficient older machines were installed in the United States, bringing the nation's total capacity to approximately 2,400 MW. 1999 saw wind capacity jump 40.8 percent over the previous year.

An important catalyst to this unprecedented growth was the expiration of the wind energy production tax credit in June 1999. Developers raced to complete projects before the expiration deadline. The tax credit has since been retroactively reinstated and will run through Dec. 31, 2001. Wind energy producers can continue to receive inflation-adjusted 1.5 cents/kWh tax credits for utility-scale projects. AWEA expects the continuing tax credit to spur more growth in the wind energy market.

Top Wind Markets	1998 MW Additions	1998 MW Year-end Total	1999 MW Additions*	1999 MW Year-end total
Germany	793	2,872	1,200	4,072
United States	193	1,770	732	2,502
Denmark	310	1,433	300	1,733
Spain	368	822	650	1,722
India	82	1,015	62	1,077
United Kingdom	10	334	18	534
Netherlands	50	375	53	428
China	55	224	76	300
Italy	94	199	50	249
Sweden	54	176	40	216

**Values for U.S., UK, China, India and Netherlands are final. Estimates for Germany, Spain and Denmark are confirmed. Italy and Sweden are unconfirmed estimates. Additions only include projects that are installed and operating by the end of the calendar year. U.S. figures are net of retired projects.*

Source: American Wind Energy Association

Table 6-3 *Top Wind Markets*

Other driving forces included progressive state policies, especially in Minnesota and Iowa, and the movement toward customer choice and green power programs in several states. Wind energy's relatively low cost has led more than 80 utilities to offer a wind energy-based product to customers.

"After struggling for most of the '90s, it (wind energy) has come of age at the very end of the millennium," states the AWEA report. "One reason for the limited activity in the mid-'90s was the uncertainty caused by deregulation of the electric sector, which caused many utilities to re-evaluate their priorities, and freeze any new investment in new capacity. In the meantime, wind technology has continued to mature, gradually convincing the electric industry that it is ready for broad deployment."

The highest levels of development activity in the next few years are expected to be concentrated in the plains states and in Texas, which has mandated 2 GW of new renewable capacity in the coming decade. In the Northeastern region, restructuring legislation is opening the market to green power producers.

Merchant Potential

Enron Wind Corp. recently dedicated its Green Power I wind power facility near Palm Springs, CA. The 22-turbine, 16.5 MW project was built solely to supply emerging green power markets and is the first major renewable power plant to enter California's market since it opened to competition in 1998.

Green Power I began producing power in June 1999. Traditionally, wind power has been sold only under long-term contract to utilities; however, the Green Power I facility was built without contract and its power is being sold through retail marketers.

The facility was developed, constructed, and is operated by Enron Wind Corp. The project uses advanced Zond Z-750kW Series wind turbines. With 158 and 164 foot rotor diameters, approximately the size of the wingspan of a MD-11 jumbo jet, the Z-750kW wind turbines are the largest manufactured in the United States.

New Wind Technologies

The U.S. Department of Energy (DOE) has been working with the nation's wind turbine industry to improve technology and lower costs since 1992. The first turbines created under these partnerships are already on the market, and a whole new generation of turbines is expected to arrive in 2002.

Two new turbines are under development. In 1994, DOE announced a $40 million program to develop a new generation of innovative utility wind turbines. The cost-effective turbines are expected to expand markets for U.S. companies in both the United States and in Europe, where competition for new wind projects is driving down costs.

Eight industry teams created concepts for new utility wind turbine rates up to 1 MW. In 1996, the National Renewable Energy Laboratory selected two firms, Zond Energy Systems Inc. and The Wind Turbine Co., to move forward with their concepts.

Zond, a subsidiary of Enron Wind Corp., is developing the A-56, which will probably be a 1 MW machine. Its architecture has not yet been determined, but it may use a direct-drive generator alone or in combination with a conventional gearbox. Significant departures from conventional design are expected, including purpose-designed airfoils and low-solidity, flexible blades with individual pitch control. Taller, low-stiffness towers are expected, as are advanced control strategies to optimize energy capture and reduce loads.

The Wind Turbine Co. is designing the WTC 1000, a lightweight, two-bladed, downwind machine. The megawatt-scale turbine will include purpose-designed blades with individual pitch control, a variable coning rotor, highly integrated structure and drivetrain, load-mitigating control strategies, simplified fluid systems, and an extremely tall guyed tower. The WTC is targeted for applications in midwestern states.

DOE is also working with three small turbine manufacturers, selected through competitive solicitation, to improve their turbines. The goal is to develop tested systems up to 40 kW in size that achieve a cost/performance ratio of 60 cents/kWh at sites with annual average wind speeds of at least 12 miles per hour. Cost/performance ratio is defined as the initial capital cost of the turbine divided by its annual energy capture.

Bergey Windpower Co. is working to improve cost/performance ratio for its BWC Excel 40 by designing a turbine with minimal maintenance requirements. The BWC Excel 40 is a 40 kW turbine targeted for battery charging in the village power market. It is a three-bladed, upwind, variable-speed machine with a direct-drive permanent-magnet alternator. Rotor blades will be pultruded fiberglass in three lengths for use in different wind regimes. The guyed lattice towers will be available in three heights. Project cost/performance ratio is 38 cents/kWh. DOE is funding $1.21 million of the research.

WindLite Corp. is developing an 8 kW, variable-speed, direct-drive machine with a rotor diameter of 23 feet. The turbine uses a wound-rotor generator and proprietary controller that significantly increases its battery-charging efficiency compared to permanent-magnet generators. The projected cost/performance ratio for the WLC 7.5 is 46 cents/kWh. DOE is providing $1.43 million in funding.

World Power Technology makes six small turbine models. Its Windfarmer, a 7 kW battery-charging wind turbine, is a three-bladed, upwind, variable-speed machine using a direct-drive, permanent-magnet generator. Fiberglass blades will be used on a 16-foot diameter rotor. The machines will use a unique, patented angle-furling governor for protection in high winds. World Power is also developing a counter-weighted, tilt-down 90-foot tower. Project cost/performance ratio is 59 cents/kWh. DOE is providing $1.25 million in funding.

Siting Hurdles

Despite its status as the fastest growing renewable energy source, wind power faces numerous obstacles in siting and permitting. This is true for both large projects being built to sell power to utilities and small projects being built for a single user. Regulations and laws governing power project siting are becoming ever more complex, and state and federal siting agencies are not as likely to approve power projects without extensive review. Various interest groups have become more involved in siting procedures as well. (Figure 6-2)

***Fig, 6-2** Mitsubishi 600 kW turbines in Tehachapi, CA. More than 5,000 wind turbines are sited in this mountainous area, showcasing old and new technologies.*

Large wind projects raise many of the same siting issues as other energy projects. There may be concern about truck traffic during construction, health effects of electromagnetic fields from transmission lines, and social issues. Wind projects also face some unique challenges that require special consideration.

Unique Considerations

Visual and noise impacts must be addressed. Wind turbines are highly visible structures and they often need to be sited in conspicuous locations, such as on ridges or hillsides. They also generate noise that can be bothersome to area residents. Those problems can be mitigated through noise abatement, design adjustments, and other measures.

Recent design improvements have greatly decreased the noise generated by today's wind turbines. (Table 6-4)

Source/Distance	A-weighted Sound level (dB)
Civil defense siren	140-130
Jet takeoff/200 feet	120
Rock music concert	110
Pile drive/50 feet	100
Ambulance siren/100 feet	90
Boiler room	90
Pneumatic drill/50 feet	80
Printing press	80
Kitchen garbage disposal	80
Freeway/100 feet	70
Vacuum cleaner/100 feet	60
Department store/office	60
Wind turbine/250 feet	50-55
Light traffic/100 feet	50
Private business office	50
Large Transformer/200 feet	40
Soft whisper/5 feet	30
Threshold of hearing	0-10

Sound is typically measured in decibels. The decibel scale is logarithmic. Outside of a laboratory, a 3 dB change is barely discernable, a change of 5 dB will likely result in a noticeable community response, and a 10 dB increase is heard as an approximate doubling in loudness.

Source: NWCC and Danish Wind Turbine Manufacturers Assoc.

Table 6-4 *Noise Comparisons*

Impacts on birds and other local wildlife must also be considered. In some locations, wind turbines and their ancillary equipment have killed raptors, such as hawks and eagles. Pre- and post-construction studies may be necessary to measure the project's impact on wildlife and to create strategies for offsetting them. Soil erosion is another potential problem that may be addressed during the siting process.

Landowner's rights may also be an issue in wind project siting. Wind plants often pay rents or royalties for land use, which can be a benefit to landowners, but it may also raise concerns. A turbine on one resident's land may interfere with a neighbor's ability to develop a wind project.

Guidelines

Successful wind project siting depends on negotiation to balance concerns and benefits. Details vary widely from site to site, but the National Wind Coordinating Committee (NWCC) recommends a few guidelines, including significant public involvement, reasonable time frames, clear decision criteria, coordinated siting processes, expedited judicial review and advance site planning. (Tables 6-5 and 6-6)

Early public involvement allows the public to have its interests factored in early in the siting process. Without this, there is a much greater likelihood of later opposition and costly litigation. The public, particularly residents living near a proposed site, should be notified of the siting application, and the siting agency should hold public meetings and accept public comments.

Open siting processes with long delays are a legitimate concern for wind developments. Establishing reasonable time frames for review of applications, hearings, and a final decision from the siting agency is one way to avoid unnecessary delays.

The siting agency should make the criteria for its decisions clear at the beginning. The agency should list all the factors to be considered, specify how the factors are weighed against each other, and set minimum requirements to be met by the project. The factors will vary depending on the circumstances.

The NWCC notes eight key elements for the development of a successful process for permitting wind energy facilities:

Significant public involvement,
Issue-oriented process,
Clear decision criteria,
Coordinated permitting process,
Reasonable time frames,
Advance planning,
Efficient administrative and judicial review
Active compliance monitoring.

Source: NWCC

Table 6-5 *Principles of Success*

1. Land use
2. Noise
3. Birds and other biological resources
4. Visual resources
5. Soil erosion and water quality
6. Public health and safety
7. Cultural and paleontological resources
8. Socioeconomics, public services and infrastructure
9. Solid and hazardous wastes
10. Air quality and climate

Source: NWCC

Table 6-6 *Top 10 Permitting Issues*

Land Use

Unlike most power plants, wind generation projects are not land intensive. On a MW output basis, the land required for a wind project exceeds the amount of land required for most other energy technology, but the physical project footprint covers only a small portion of that land. For example, a 50 MW wind facility may occupy a 1,500-acre site, but it will only use three to five percent of the total acreage, leaving the remainder available for other uses.

Because wind generation is limited to areas with strong and fairly consistent wind resources, most wind generation is sited in rural and relatively open areas that are often already used for agriculture, grazing, recreation, forest management, or seasonal flood storage.

To ensure that a wind project is compatible with existing land uses, the layout and design of the wind project can be adjusted in a variety of ways, including the following:

- selecting equipment with minimal guy wires,
- placing electrical collection lines underground,
- placing maintenance facilities off site,
- consolidating equipment on the turbine tower or foundation,
- consolidating structures within a selected area,
- using the most efficient or largest turbines to minimize the number of turbines required,
- increasing turbine spacing to reduce density of machines,
- using roadless construction and maintenance techniques, or
- using existing access roads.

Other land use strategies include buffer zones and setbacks to separate wind projects from sensitive or incompatible land uses. Land use agencies in California have established setbacks ranging from two to four times the height of a turbine or a minimum of 500 to 1,200 feet from any residential area. Minnesota has established minimum setbacks of 500 feet from occupied dwellings.

For the Birds

The problem of birds, especially raptors, flying into wind turbines has been the most controversial biological consideration affecting wind siting. Wind developments have produced enough bird collisions and deaths to raise concern from wildlife agencies and conservation groups. On the other hand, some large wind facilities have been operating for years with only minor impacts on birds. Smokestacks and radio and television towers have actually been associated with much larger numbers of bird deaths than wind facilities have, and highways and pollution account for a great many as well.

Whether or not this becomes a serious sighting issue tends to depend on the protective status or number of bird species involved. Most raptors are protected by state and federal laws and any threat to them may cause siting concerns.

Both the wind industry and government agencies are sponsoring or conducting research into this problem. Studies are underway comparing mortality at lattice and tubular towers and investigating birds' sensory physiology and how it affects their ability to detect components of wind turbines. One study is painting colors on turbine blades to observe birds' reactions.

Wind farms are thought to affect wildlife in several other ways, including the following:

- direct loss of habitat,
- indirect habitat loss from increased human presence, noise or motion of operating turbines,
- habitat alteration resulting from soil erosion or construction of obstacles to migration,
- collision with structures, turbine blades or wires, and
- electrocution from contact with live electrical wires.

The NWCC recommends several strategies for dealing with biological resource siting issues—consultation, surveys, and risk reduction.

Planning and coordination with permitting agencies can reduce the chances of project delays. Most permitting agencies recommend that wind developers consult with them and appropriate natural resource protection

Big Spring Keeps on Turning

As early as 1993, TXU Electric and Gas investigated the level of demand for renewable energy in Texas. Encouraged by the enthusiasm of its customers toward green energy, TXU unveiled plans for the $40 million Big Spring wind power project near Midland, TX in December 1998. Developed by York Research, the project has 46 turbines with a total capacity of 34 MW. The final phase, completed in April 1999, saw the commissioning of the largest commercial wind turbines in the world—four Vestas V66 turbines standing approximately 260 feet tall above the elevated plateau of west Texas ranch land.

TXU believes that the project is testament to the fact that as power technologies advance, electricity generated by renewable resources will become more common and economic.

The Big Spring project is built on mesas, rising 195 to 295 feet above the surrounding areas. The winds accelerate as they move up over these mesas. Annual average hub-height wind speeds range from 18.4 to 22.2 mph over the site.

There are three phases to the site. Phase I has 16 Vestas V47 660 kW turbines, Phase II has 26 Vestas V47s, and Phase III has the four Vestas V66 1,650 kW turbines.

Projected annual electricity generation for Big Spring is 117 million kWh.

Both turbine models use three rotor blades of epoxy and fiberglass composite. Crosswind separation of the machines is nominally 3.5 rotor diameters. Row-to-row spacing of the machines exceeds 10 rotor diameters to minimize the impact of turbulence from adjacent rotors.

The turbine control system monitors turbine starts and stops under normal operating conditions and also protects the turbines under extreme emergency conditions such as faults caused by a loss of grid load while under power or a component failure. In addition, the system manages the power output of each turbine by pitching the blades and changing the generator slip to maximize energy production while minimizing loads at wind speeds greater than 31 mph.

The control system is operated by a digital computer using Vestas-developed programs. Portions of the system are located in the base of the tower and in the nacelle of the wind turbine. These are linked by fiber optic lines to minimize interference and damage from lightning.

A key feature of the control system is OptiSlip, which controls loads and spikes from the turbines under high wind speeds. OptiSlip allows the turbine to operate in a similar way to a variable speed machine, preventing the drive line of the machine from experiencing torque spikes.

Fig. 6-3 *TXU's Big Spring wind project has 46 turbines for 34 MW of capacity.*

agencies early in the site selection process to determine the potential for conflicts. It is important to find out whether protected plants and animals inhabit, use, or migrate through the area. Unique or rare habitat types, such as savannas, can raise interest and alternative sites may be needed.

Biological surveys can be helpful, but the timing is important. Some necessary information can only be obtained at a certain time of the year. Protected plants may only bloom for a few weeks or months at a time and bird use or migration patterns may need to be studied over several seasons or years.

Equipment selection can reduce the risk of high bird mortality, but the best plan is to avoid sites near major bird feeding, roosting, or resting areas. Research is ongoing, but to date there are no designs or modifications that have been statistically proven to significantly reduce the risk of bird collisions. Unless protected plants or animals are involved, most permitting agencies tend to find the non-collision effects of wind development on wildlife to be insignificant.

Fig. 6-4 *A 600 kW Tacke wind energy turbine. It features a 200-foot-tall tower and 75-foot blades.*

Visionary Planning

Visual or aesthetic concerns are also a common issue in wind siting. Wind projects tend to be located in rural or remote areas with few area residential developments. Potential for visual impact is sometimes considered as part of the evaluation of land use, and the degree to which the visual quality of a project is addressed will vary. Elements that can influence the visual impact of a wind project include spacing, design and uniformity of the turbines, markings on turbines and other structures, spacing of turbines, design and uniformity of the turbines, roads built on slopes, and service buildings.

There is considerable motion in turbine blades and this motion is intensified when the turbines are placed close together, are of different designs, or rotate in different directions. Adequate spacing between turbines and between rows or tiers of turbines mitigates visual impact.

When turbines are sited on ridgelines, the units are visible for greater distances. Against the sloping terrain, surfaces exposed by construction of access roads and turbine pads may contrast with existing soils and vegetation. From a distance, the visual impact of the roads may be greater than that of the turbines. Constructing roads on ridges also may increase erosion. (Fig. 6-4)

It is generally recommended that developers contact any agencies with jurisdiction for any maps, plans, guidelines or design standards in that particular area. Design strategies can be used to reduce the visual impact, including the following:

- using the local landscape to minimize visibility of access and service roads and to protect soils from erosion,
- consolidation of roads or use of grading over vegetation for temporary access without road construction,
- use of low-profile building designs,
- use of uniform color, structure types, and surface finishes,
- consolidating electrical lines and roads into a single right of way or corridor,
- limiting the size, color, and number of labels on turbines,
- limiting size and number of advertising signs on fences and facilities, and
- using air lift for transport of turbine components and installation.

SOLAR

Each day more solar energy falls to the earth than the total amount of energy the planet's 6 billion inhabitants would consume in 26 years. While it's neither possible nor necessary to use more than a small portion of this energy, we've hardly begun to tap the potential of solar energy.

Although every location on earth receives sunlight, the amount received varies greatly depending on geographical location, time of day, season, and light. The southwestern United States is one of the world's best areas for sunlight. This desert region receives almost twice the sunlight of other regions of the country. Solar energy systems use ether solar cells or some form of solar collector to generate electricity or heat for homes and buildings. The primary solar energy technologies for power generation are photovoltaics and thermal systems. (Fig. 6-5)

Fig. 6-5 *An array of thin glass mirror tiles attached to a stainless steel stretched membrane shows off heliostat technology from the U.S. Department of Energy and the National Renewable Energy Laboratory.*

Photovoltaic devices use semiconductor material to directly convert sunlight into electricity. Solar cells have no moving parts. Power is produced when sunlight strikes the semiconductor material and creates an electric current. Solar cells are used to power remote homes, satellites, highway signs, water pumps, communication stations, navigation buoys, street lights, and calculators.

Solar thermal systems generate electricity with heat. Concentrating solar collectors use mirrors and lenses to concentrate and focus sunlight onto a receiver mounted at the system's focal point. The receiver absorbs and converts sunlight into heat. The heat is then transported to a steam generator or engine where it is converted into electricity.

Solar energy technologies offer a clean, renewable, and domestic energy source. The generating systems they power are also modular so they can be constructed to meet any size requirement and are easily enlarged to meet changing energy needs.

Solar energy technologies have made huge technological and cost improvements, but except for certain niche markets like remote power applications, they are still more expensive than traditional energy sources. Researchers continue to develop technologies that will make solar energy technologies more cost competitive.

Virginia Industrial Park Boasts 42 kW Solar Array

PowerLight Corp. installed the largest roof-integrated, thin-film solar electric system in North America in 2000. The project is sited in an eco-industrial park at the Port of Cape Charles, VA, and consists of 10,000 square feet of PowerGuard roofing tiles. "PowerGuard tiles are a revolutionary concept in building architecture," said Dan Shugar, PowerLight executive vice president. "In addition to generating solar electricity, the tiles insulate the building, reducing the cost of heating and air-conditioning, while also protecting and extending the life of the roof. They're an integral part of the rooftop, joined by a tongue-and-groove design that requires no roof penetration or adhesives, thus eliminating leakage and related maintenance."

The system uses a newly commercialized thin-film solar electric module called the Millennia, from BP Solarex.

The 42 kW solar array is mounted on a building n the Cape Charles Sustainable Technology Industrial Park, a public-private initiative funded in part by the National Oceanic and Atmospheric Administration, the DOE and the Virginia Department of Environmental Quality. Funding for the PowerLight system was provided in part by the Virginia Alliance for Solar Electricity and the Utility PhotoVoltaic Group's TEAM-UP program.

BIOMASS

Biomass energy is derived from the energy stored in plants and organic matter. It is used to meet a variety of energy needs including generating electricity, heating homes, fueling vehicles, and providing process heat for industrial facilities.

Biomass also can be converted into transportation fuels such as ethanol, methanol, biodiesel, and additives for reformulated gasoline. Biofuels are used in pure form or blended with gasoline. Typical biomass fuels include the following:

Ethanol – Ethanol, the most widely used biofuel, is made by fermenting biomass in a process similar to brewing beer. Currently, most of the 1.5 billion gallons of ethanol used in the United States each year is made from corn and blended with gasoline to improve vehicle performance and reduce air pollution.

Methanol – Biomass-derived methanol is produced through gasification. The biomass is converted into a synthesis gas (syngas) that is processed into methanol. Of the 1.2 billion gallons of methanol annually produced in the United States, most is made from natural gas and used as a solvent, antifreeze, or to synthesize other chemicals. About 38% is used for transportation as a blend or in reformulated gasoline.

Biodiesel – Biodiesel fuel, made from oils and fats found in microalgae and other plants, can be substituted for, or blended with diesel fuel.

Reformulated gasoline components – Biomass can also be used to produce reformulated gasoline components such as methyl tertiary butyl ether (MTBE) or ethyl tertiary butyl ether (ETBE).

Biomass resources include wood and wood wastes, agricultural crops and their waste byproducts, municipal solid waste, animal wastes, waste from food processing, and aquatic plants and algae. The majority of biomass energy is produced from wood and wood wastes, followed by municipal solid waste, agricultural waste, and landfill gases. **Dedicated energy crops** – fast growing grasses and trees grown specifically for energy production – are also expected to make a significant contribution in the next few years.

Landfill gas plants collect methane gas – the primary component of natural gas – to run generators. Methane is a flammable gas produced from landfill wastes through anaerobic digestion, gasification, or natural decay. More than 100 power plants in 31 states burn landfill-generated methane.

Wood-related industries and homeowners consume the most biomass energy. The lumber, pulp and paper industries burn their own wood wastes in large furnaces to heat boilers to supply energy needed to run factories. Homeowners burn wood in stoves and fireplaces to cook meals and warm their residences. Wood is the primary heating fuel in 3 million homes and is used to some degree in 20 million homes.

Landfill Gas-to-Energy Powers Iowa

At Metro Park East Sanitary Landfill near Des Moines, Iowa, a 6.4 MW landfill gas-to-energy project is paving the way for a nationwide Waste Management Corp. initiative. Waste Management has more than 60 Caterpillar generator sets running on landfill gas and providing more than 48 MW of electricity to utilities across the country.

The Metro Park site was installed to beta test new technologies for use at other installations. With some of the most advanced technology, the facility has logged more than 99% up-time, including scheduled maintenance on eight Caterpillar G3516 SITA-LE generator sets.

Each genset is capable of producing 800 kW of continuous power. Since landfill gas can fluctuate from the normal 55% methane content, electronic control modules monitor the amount and quality of the intake fuel, optimizing the engines' performance.

Using advanced methane gathering methods, the landfill produces more than 3.2 million cubic feet of gas per day. An underground piping system spans 5.5 miles, connecting 70 eight-inch diameter well bores drilled to a depth of 86 feet.

The electricity from this site is sold to the local utility for grid distribution.

The facility burns the equivalent of 112,000 barrels of oil annually in landfill gas. The methane gas, which contributes to ozone depletion when not properly disposed of, is effectively burned off in the on-site power plant.

Fig. 6-6 *Waste Management has installed more than 60 Caterpillar G3516 SITA-LE generator sets at landfills across the country.*

Biomass energy is used to make electricity, liquid fuels, gaseous fuels, and a variety of useful chemicals. Because the energy in biomass is less concentrated than the energy in fossil fuels, new technologies are required to make this energy resource competitive with coal, oil, and natural gas. Industry and agriculture need superior energy crops and cost-effective conversion technologies to expand the use of renewable biomass.

Biomass energy generates far fewer air emissions than fossil fuels, reduces the amount of waste sent to landfills, and decreases reliance on foreign oil. To increase use of biomass energy, dedicated energy crops must be developed, system efficiencies must improve, infrastructure to efficiently transport biofuels must be developed, and the cost of biomass energy must become more cost-competitive with fossil fuels.

GEOTHERMAL ENERGY

Miles beneath the earth's surface lies one of the world's largest energy resources – geothermal. Our ancestors used geothermal energy for cooking and bathing since prehistoric times. Today, this enormous energy reservoir supplies millions of people with clean, low cost electricity.

Geothermal energy is the heat contained below the earth's crust. This heat – brought to the surface as steam or hot water – is created when water flows through heated, permeable rock. It's used directly for space heating in homes and buildings or converted to electricity.

Most of the country's geothermal resources are located in the western United States. The United States has more than 2,700 MW of geothermal electric power capacity, coming from naturally occurring steam and hot water. The only scratches the surface of the potential electricity production of geothermal resources.

Geothermal resources come in five forms:

- hydrothermal fluids
- hot dry rock
- geopressured brines
- magma
- ambient ground heat

Of these five, only hydrothermal fluids have been developed commercially for power generation. Three technologies can be used to convert hydrothermal fluids to electricity. The type of conversion used depends on whether the fluid is steam or water, and its temperature.

Steam – Conventional steam turbines are used with hydrothermal fluids that are wholly or primarily steam. The steam is routed directly to the turbine, which drives an electric generator, eliminating the need for the boilers and conventional fuels to heat the water.
High-temperature water – For hydrothermal fluids above 400 degrees F that are primarily water, flash steam technology is usually used. In these systems, the fluid is sprayed into a tank held at a much lower pressure than the fluid, causing some of the fluid to rapidly vaporize, or flash, to steam. The steam is used to drive a turbine, which again, drives a generator. Some liquid remains in the tank after the fluid is flashed to steam, if it's still hot enough, this remaining liquid can be flashed again a second tank to extract even more energy for power generation.
Moderate-temperature water – For water with temperatures less than 400 degrees F, binary-cycle technology is generally most cost effective. In these systems, the hot geothermal fluid vaporizes a secondary – or working – fluid, which then drives a turbine and generator.

Steam resources are the easiest to use, but they are rare. The only steam field in the United States that is commercially developed, the Geysers, is in northern California. The Geysers began producing electricity in 1960. It was the first source of geothermal power in the country and is still the largest single source of geothermal power in the world.

Hot water plants, using high- or moderate-temperature geothermal fluids, are a relatively recent development. However, hot water resources are much more common than steam. Hot water plants are now the major source of geothermal power in both the United States and the world.

Today's hydrothermal power plants with modern emissions controls have minimal impact on the environment. The plants release little or no carbon dioxide. Geothermal power plants are very reliable when compared to conventional power plants. For example, new steam plants at the Geysers are operable more than 99% of the time.

In some parts of the world, geothermal systems are cost competitive with conventional energy sources. It is anticipated that as technology improves, the cost of generating geothermal energy will decrease.

End Notes

1. "Wind Energy Issue Brief No. 3," National Wind Coordinating Committee, www.nationalwind.org.
2. "Permitting of Wind Energy Facilities," National Wind Coordinating Committee.

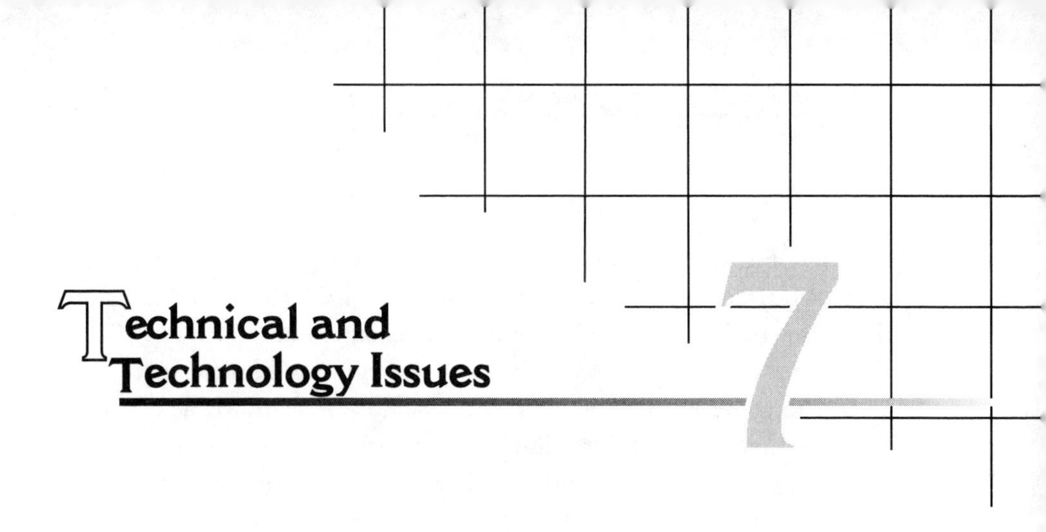

7 Technical and Technology Issues

When you get right down to it, technology is actually the big question when looking at the future of the electric power industry with regard to just about any industry trend or niche. Distributed generation is certainly no exception. Technological advances and acceptance are crucial for distributed generation's survival.

Many of the technologies discussed here are, or soon will be, commercially available. But they are generally still too expensive for common uses. With each generation of technology that passes, their costs come closer to competing with traditional fossil and nuclear generation. Other factors affect their competitiveness as well, including governmental incentives, cost of the various fossil fuels, availability of net metering, and many others.

Regulations Affect Fuels, Costs

Potential effects of future restrictions on CO_2, NO_x and SO_2 include significant shifts in the fuels used to generate electricity, leading to a coal decline from more than 50 percent today, to less than 10 percent by 2020, according to a two-year study by EPRI.

The "Energy-Environment Policy Integration and Coordination" (E-EPIC) study details implications of emissions control, air quality, and global climate change policies affecting the U.S. energy sector.

Over the next 20 years, natural gas generation would increase from 15 percent to about 60 percent. This shift would cause many viable coal-fired generating plants to fall into disuse, thereby stranding investments in not only the plants, but also recently installed emission control equipment.

During the same period, a national investment of an estimated $160 billion in new gas-fired generating plants would be needed to meet domestic electricity demands. The study found, moreover, that the significant increases in the production and transportation of natural gas could prove difficult or even infeasible.

"Our study revealed that even if this initial shift from coal to gas could be accomplished, reliance on natural gas for electricity generation could not be maintained at such a high level for an extended period of time," said Dr. Gordon Hester, EPRI E-EPIC manager. The study projects that increasing demand would cause natural gas prices to rise, prompting shifts to electricity generating technologies that use less costly fuels while producing lower emissions. As a result, natural gas used for electricity generation would decline significantly after 2025. After that time, the study indicates electricity will likely be generated by a mix of biomass, wind, natural gas, and large amounts of new and cleaner coal-fired generating technologies. The study suggests that new nuclear power plants also have the potential to play a significant role in future electricity generation.

The E-EPIC study indicates strong ties between potential shifts in electricity generation sources and the U.S. economy. Under the current policy direction, the price of electricity is projected to increase 50 percent (in real dollars) by 2020, affecting industrial, commercial and residential electricity consumption. The economies of coal-producing regions would be especially hard hit, and pressures on the energy sector would likely produce a ripple effect in the general economy. Between 2005 and 2020, the average annual increase in overall consumer prices attributable to the emission restrictions would be more than 2 percent.

The study concludes that synchronizing emission reductions with technological advances is crucial to meeting economic and environmental goals.

Such coordination could enable the U.S. energy system to contribute more efficiently to the long-term stabilization of atmospheric carbon levels. The current policy direction would require substantial reductions in emissions before the deployment of advanced technologies could allow a transition to a sustainable, low carbon-emitting energy system.

"One way to avoid this gap in timing is to extend the schedule for emission reductions – to wait for technology to catch up," says Hester. "Another way is to accelerate the pace of technology development through an aggressive R&D campaign so that technology can catch up more quickly."

Standard 1008

On-site standby power solutions for residences are relatively new for most people, as are the transfer switches that provide the critical link between the utility and backup power sources. However, the residential market is growing rapidly, and more generator sets and their accompanying transfer switches are finding their way onto the shelves of megastores – and into the minds of contractors and municipal electrical inspectors throughout the United States.

Unfortunately, a considerable number of products that are being called transfer switches have never been tested to the stringent criteria contained in UL Standard 1008 – the only UL standard dedicated exclusively to transfer switches. The problem is compounded by the fact that these same switches probably display a UL label prominently on their packaging and on the switch enclosure. While this may imply compliance with an applicable UL standard, this is often not the case. This leaves contractors and electrical inspectors in the compromised position of approving the installation of a transfer switch that appears acceptable when in fact it may not be. It could also be argued that installation of transfer switches that have not complied with UL1008 is a violation of the National Electric Code that requires that products be tested and listed only for their intended use.

Unless a transfer switch is evaluated and complies with the criteria in UL1008, it can be wholly unfit for service. When pressed into the service it was ostensibly intended to perform, these switches can destroy the generator to which they are connected, cause a fire, or both. The potential liability issues arising from this situation are enough to raise goose bumps on the arms of inspectors and contractors everywhere.

IEEE Standards

An IEEE-sponsored working group has developed an interconnection standard that will simplify the process of interconnecting photovoltaic systems (PV) with an electric utility. This standard is the first IEEE standard of its kind for allowing utility interconnections of non-utility-owned distributed generation equipment. The unique aspects of this standard include tightly defined requirements for the interconnecting hardware that can be tested by an independent test laboratory such as Underwriters Laboratories, which is expected to remove former barriers to PV use throughout the country.

John Stevens of Sandia National Laboratories chaired the working group, which included about 25 members representing the utility industry, the PV industry, PV inverter manufacturers and PV researchers. The effort was sponsored by IEEE Standards Coordinating Committee 21 (SCC21). It required a little over three years from initial announcement of the project to final approval by the IEEE Standards Board. PV interconnection hardware can now be designed to meet the standard, eliminating the need for specialized hardware for different utility jurisdictions. The standard includes very specific requirements for systems of up to 10kW, but it covers systems of all sizes. The IEEE PV interconnection standard, identified as IEEE Std 929-2000, is known informally as IEEE 929.

The standard actually applies to the PV inverter, the device that converts the PV dc energy into utility-compatible ac energy. Similar inverters are used in other distributed generation systems such as fuel cells and microturbines. Many of the requirements for interconnection that are described in IEEE 929 might also be adopted for these other technologies.

IEEE 929 provides guidance for operating voltage and frequency windows, trip times for excursions outside these windows, requirements for waveform distortion, as well as defining a non-islanding inverter. A parallel effort was performed at Underwriters Laboratories where a test procedure,

UL 1741, was written to verify that an inverter meets the requirements of IEEE 929.

Technology Initiatives

EPRI has announced three new Strategic Science & Technology initiatives to help achieve the key destinations defined in its Electricity Technology Roadmap.

The Advanced Coal Initiative is intended to help resolve the energy/carbon conflict and manage the global sustainability challenge. Efforts will focus on development of ultralow-NO_x combustion processes to optimize energy conversion efficiency while minimizing operational impacts; conceptual design of cost-competitive, ultrahigh efficiency gasification plants with near-zero CO_2 emissions; and investigation of environmental control technologies and combustion processes for cost-effective NO_x, SO_x and air toxic reductions.

The second initiative focuses on design of new retail products, focusing on customer-managed service networks with a goal of boosting their economic productivity and prosperity. A consumer research approach will be applied to guide development and launch of energy-related products and services that integrate advanced technology with targeted pricing strategies to fulfill consumer needs within specific market niches.

The third initiative, dubbed the Grid-Connected Hybrid Vehicle Initiative, is intended to build on the EPRI's Transportation Initiative, which was completed in 1999. That initiative helped advance battery technology for electric and hybrid vehicle applications. The new initiative's analyses of costs, fuel economy, emissions performance, customer acceptance, commercialization, and other issues for hybrid systems and components will assist development of market-ready vehicles offering global energy efficiency and environmental benefits.

Nine other strategic initiatives are already under way:

Future of Power Delivery—Development of transmission, distribution, security and communications hardware and software necessary for highly reliable, cost-effective operation of the next-generation power grid. Seeks to maximize utilization of existing transmission assets and rights-of-way by accelerating commercial development of technologies such as superconducting power cables.

Advanced Power Electronics—Development of advanced silicon and wide-bandgap semiconductor technologies for rapid, reliable operation, maximum asset utilization and optimal power quality in complex energy circuits.

Complex Interactive Networks/Systems—Establishment of adaptive, self-healing models and control strategies for the stable and secure operation of the next-generation power grid.

Power Market Redesign—Theoretical development, experimental testing and demonstration of a new market design coupling power and transmission rights trading and avoiding inefficiencies and inequities of poorly designed market structures.

Low Pressure Turbine Performance—Development, testing and demonstration of physical phenomena and techniques for substantially improving thermal efficiency of low pressure steam turbines in fossil and nuclear power plants.

Advanced Nuclear—Investigation of advanced reactor technologies, advanced plant information systems, long-term equipment performance and waste disposal issues.

Human Performance—Development of capabilities and tools for effective human performance management by monitoring, anticipating and addressing factors with adverse impact on human reliability and productivity in specific workplace environments.

Climate Adaptation—Scientific and economic assessment of adaptation as an alternative risk management option for adjusting economic, social and ecological systems to reduce their vulnerability to potential climate change.

Materials Science—Identification, investigation, and development of condition and remaining life assessment techniques, damage mitigation methods, and advanced materials and processes for power generation, transmission, and distribution systems. Work includes fundamental investigations of critical damage mechanisms such as high-temperature corrosion, as well as experimental and field tests of preventive methods such as use of regenerative biofilms for aqueous corrosion control.

Best, Worst Technologies

Net world electricity consumption should reach 22 trillion kWh by 2020, almost double its current level, according to "Natural Gas: Technologies and Applications" from Ecole Polytechnique Montreal. The North American market of Canada and the United States currently is the largest integrated market in the world with almost one-third of total world consumption. Despite predictions of weak growth, only 39 percent from 1996 to 2020, the North American market will remain dominant, accounting for one-fourth of world consumption in 2020.

This report jumps on the bandwagon with other industry reports, predicting natural gas turbines and combined-cycle systems to lead new installations, ranking it first of 120 technologies examined. The commercial potential of advanced natural gas turbine systems is expected to be $5 billion annually in the U.S. market and about $1 trillion worldwide annually between 1999 and 2010.

Other natural gas technologies tagged as having high potential include cogeneration, tricogeneration, SOFC and PEM fuel cells. (Table 7-1) Fuel cells and microturbines represent 5 percent of the world electricity market and are expected to reach 20 percent in the next five years.

The report also notes that while there are several gas technologies with strong commercial potential, they have viable rivals – especially in clean coal electricity generating technologies like IGCC and fluidized bed combustion.

Technologies of interest to electricity generators that were tagged as having the lowest potential included recovery of energy liberated by natural gas expansion and phosphoric acid fuel cells.

Technology	Time Horizon*
Advanced turbine systems	2000
High-efficiency engines for liquefied natural gas (LNG) and compressed natural gas (CNG)	2005
Tricogeneration (heat, power and refrigeration)	2000
Fast-fill refueling stations for LNG and CNG	2000
Direct CNG fuel uses	2000
Uninterrupted power supply backup systems	2000
High-pressure gas cylinders for CNG (onboard vehicles)	2000
Solid oxide fuel cells, 25 kW – 25 MW, 45+ percent efficiency	2010
Proton exchange membrane (PEM) fuel cells	2010
1 kW – 250 kW, 40+ percent efficiency	
Hybrid gas turbine/fuel cell	2015+
70+ percent efficiency and less than 2 ppm NO_x	
PEM for ground transport	2005
High-temperature heat exchangers or recuperators	2005
Gas microturbines, high efficiency	2005
Micro-cogeneration	2010
Cogeneration based on fuel cells	2005

**Year the technology is expected to be used to a significant extent.*
Source: Ecole Polytechnique

Table 7-1 *Most Important Natural Gas Technologies For Electricity Generation and Transportation*

Utility Use And Issues

8

Large utility plants can be at a disadvantage in a competitive environment. They can generate a large amount of electricity at a moderate price, but running these plants at low loads can be problematic. In addition, transmission infrastructure construction is becoming more and more of an expense and problem for utilities. For innovative power generating companies, distributed generation plants can avoid both these problems by enabling capacity to be installed right where it is needed.

When a small power generation unit is placed on-site, or very close to the facility or facilities that need the power, it can eliminate costly overbuilding of capacity and expensive transmission line construction.

Mini Merchants

The mini-merchant plant for distributed generation is a relatively new concept. It refers to a distributed generation facility that seeks to match its generating portfolio to a local or regional electricity demand profile in the most efficient and economic way. These plants are typically cogeneration facilities, with overall thermal efficiencies as high as 88%. When compared directly to the separate production of electricity and thermal energy, these plants can reduce CO_2 emissions by 50% for the same amount of useful energy. They also may reduce the amount of fuel used by up to 50%.

The success of the mini-merchant plant model hinges on overall economics and how cogeneration and distributed generation fit together. For distributed generation merchant facilities to work well, several characteristics must exist: flexible dispatch, load following, duty cycle, cogeneration, power production, and defined service territory. These plants can be run on internal combustion engines or gas turbines. Some utilities and IPPs are adapting to today's market by building these sorts of facilities to enhance their generation portfolios.

The production capacity must be capable of being dispatched, cycling on and off based on the price of alternative sources of electricity. To facilitate dispatch, the mini-merchant responds to base load, intermediate load, and peak load demand requirements. Effective dispatch requires that all engines be capable of starting and synchronizing in less than 30 seconds. In most cases, this capability will be unnecessary, but could be required.

Rapid load changes must be accommodated without tripping off the load and maintenance should not be affected by repeated starting and stopping of the units. These abilities make these small plants far more flexible than standard utility-scale units.

For distributed generation applications, load-following capabilities are essential. Reciprocating engine efficiency is reasonable flat – between 40% and 100% load for individual generators. Several engines make it possible to load follow a local area from base to peak with little effect on efficiency. Large-scale utility plants do not enjoy this luxury. They generally have limited load range for top efficiency.

The difference between base load and peak averages 100%. For instance, electricity load in the summer months is low at night, when many industrial customers are closed and air conditioners are running very little. During the day, when industrial customers are running and air conditioners are cycling, power demand jumps 100% or more.

To minimize the capital cost for a distributed generation plant, it is important to match the generating equipment type to the expected duty. Peaking requirements are met through peaking generating equipment, intermediate generation is used to intermediate needs, and base load equipment provides for base load needs.

Thermal energy production – cogeneration – helps optimize efficiency for distributed generation facilities. Thermal energy production must be reliable (with or without electricity production) for this ability to truly be an asset. Natural gas engines have a fairly high exhaust temperature – generally more than 770 degrees F – corresponding to a plant thermal capacity of more than 24 MW. Heat is recovered from exhaust gases and satisfies thermal needs of the facility.

The amount of electricity produced at a small cogeneration plant or mini-merchant is determined by the size of the thermal host. This ensures that production efficiency is maintained at an optimum level. When there is little thermal need, all generation costs are absorbed on the electricity side, with none going to a thermal power cost center. If electricity is needed at a time when thermal needs are low, the decision to produce electricity versus purchasing outside electricity will depend on a comparison of the incremental cost of production and purchase. Normally, the cost of purchasing outside electricity is lowest when weather is moderate. Extremes in climate in both summer and winter increase electricity demands.

In the open market, there are times when low electricity load conditions on the grid, force utilities to discount their energy to near zero pricing. When this happens, on-site generating facilities need to have the flexibility to purchase that low-cost outside power. The goal of distributed generation, however, is the minimize reliance on the transmission grid for peaking and intermediate generation, and to produce electricity for the grid when it is economically practical.

Utility Uses

Using distributed generation resources, sited close to loads, allows utilities and other energy service providers to

- Provide peak shaving in high-load growth areas
- Avoid difficulties in permitting or gaining approval for transmission line rights-of-way

- Reduce transmission line costs and associated electrical losses
- Provide inside-the-fence cogeneration at customers' industrial or large commercial sites.

Utility Benefits

Utilities can use distributed generation as part of their overall strategy for the new competitive markets. It can be used as a subsidiary business venture, as a way to defer infrastructure expansion and expense, and more. Benefits depend on applications chosen and the needs of the particular utility.

Distributed generation can benefit utilities through

- Deferral of transmission and distribution construction
- Reduction of transmission and distribution system losses
- Balance of supply and demand
- Reductions of resource acquisition costs
- Risk management and arbitrage
- Support of ancillary services or service delivery
- Increase in service reliability
- Increase in power quality when needed.

Promote and Install

Some utilities see distributed generation as an opportunity to meet peak needs while meeting customers needs. By installing appropriate distributed generation equipment at select larger-load customer's sites, these utilities are creating customer satisfaction through personalized service and enhanced power quality and reliability while also enhancing the utility's distribution service. Properly sited distributed generation facilities can ease strain on aging, overworked utility grids. They can also reduce peak demand on those grids.

Mariah Energy purchased 126 Capstone 60 microturbines in a 2-year contract in the fall of 2000, shortly after Capstone unveiled the new turbine. Mariah Energy is a Canadian energy service provider based in Calgary, Alberta, Canada

As an original equipment manufacturer, Mariah Energy is able to resell or deploy the Capstone MicroTurbine as a component of its Heat PlusPower micro-cogeneration packages worldwide. Mariah Energy began working with the Capstone MicroTurbine system in 1999 in a landmark combined-heat-and-power (CHP) project at a commercial/residential building in Calgary. Mariah Energy's unique approach to converting the clean exhaust of the Capstone MicroTurbine maximizes the system's total efficiency to more than 80 percent, and excess power is sold back to the utility grid at premium rates.

"Almost all centralized power plant generation for the province of Alberta is coal-fired. Efficiency is about 30 percent, and a portion of that is lost in getting it to the end user. Heat generated is simply wasted, and a lot of pollutants are being generated in the process," said Paul Liddy, Mariah Energy president

"Our Capstone MicroTurbine based solution offers triple the efficiency, near-zero emissions, heat and power costs lower than utility rates and protection from outages. And with the ability to sell unused electricity back to the grid, often at a margin of several hundred percent, we can add further to a customer value proposition," Liddy said.

The Capstone MicroTurbine is a near-zero emission compact power and heat generation product deployed worldwide in a wide range of applications. The turbines are fuel-flexible and UL-listed.

The additional 126 systems will combine Mariah's exclusive heat exchangers, load management equipment and electronic controls with Capstone's ultra-low-emission microturbines. NO_x emissions are more than 10 times lower than those of other natural gas powered generators and 100 times lower than those of diesel generators. The result will be a clean "distributed micro-utility," with all equipment operated from a web-based central dispatch and control facility. The equipment is installed at no capital cost to the customer. Mariah takes responsibility for all operational and maintenance aspects, billing the end user solely for energy delivered.

"Many commercial businesses, including hotels, hospitals, and convenience stores, are looking favorably at solutions that simultaneously produce heat and power economically and efficiently," said Ake Almgren, Capstone president and CEO. "Capstone MicroTurbines are used in CHP projects in

North America, Europe, and Asia, ranging from a Walgreens drug store in Indiana to a Holiday Inn in Fargo. Mariah's state-of-the-art approach to clean, quiet, and efficient combined heat and power generation, with excess energy being sold back to the grid, is a significant step forward."

Mariah Energy Corp., a distributed micro-utility, is part of the Suncurrent Group of companies based in Calgary, Alberta, Canada. Mariah Energy, a specialist in cogeneration solutions, offers turn-key systems on a build-own-operate basis, that provide engineered heat and energy delivery for commercial, residential and industrial applications.

Capstone Turbine Corporation is a producer of low-emission microturbine systems. In 1998, Capstone was the first to offer commercial power products utilizing microturbine technology, the result of more than 10 years of focused research. Worldwide, hundreds of commercial production Capstone MicroTurbine systems are in service. (Fig. 8-1)

Supplemental Power

Commercially produced microturbines can be an economical answer to supplemental and backup power needs for today's utilities. Load in the United States grows slowly because it is a mature electric market, but it does indeed grow, necessitating power capacity additions. As discussed earlier, many utilities stopped adding capacity more than a decade ago while awaiting the outcome of deregulation.

As the guaranteed return on investment that utilities were accustomed to goes by the

Fig. 8-1 *The first Capstone 60 MicroTurbine production unit, pictured without its side panels.*

wayside with utilities old monopoly status, many utilities are looking for economical ways to add small amounts of capacity to cover peaks and short-term outages. Many are turning to the latest generation of more efficient, modular microturbines. Today's microturbines also offer relatively low emissions levels and low sound levels, making them environmentally friendly and suitable for siting in populous areas where the utilities need their power.

For example, Reliant Energy Minnegasco installed the first Capstone 60 microturbine unit at a propane and liquefied natural gas storage facility to ensure proper temperature maintenance of the stored fuels and for supplemental and backup power to the grid.

In addition, the exhaust heat of the Capstone 60 will be ported through an absorption chiller that will deliver refrigerated air, which will also be used for cooling. Reliant Energy Minnegasco already had one of Capstone's 30 kW Model 330 systems to provide load-following and standby energy to the facility's control center. The exhaust from the new system will pass through a Unifin heat exchanger for space heating at the facility.

The refrigerator-sized Capstone 60 is about the same height and width of Capstone's line of 30 kW models, and only about two feet more in depth. The Capstone 60 produces twice the amount of electricity, but does so with the same characteristics of the company's 30 kW models: only one moving assembly, no oil or other liquid lubricants, no liquid coolants and nitrogen oxide emissions of 9 parts per million without any post-combustion pollution controls.

Similarly sized natural gas generators can emit several hundred parts per million NO_x; gasoline and diesel generators can be 1,000 or more.

These kinds of turbines are being used for a variety of commercial applications, including the following:

- Hybrid electric vehicles (HEVs): onboard generation
- Resource recovery: converting oilfield and biomass waste gases into electricity
- Micro-cogeneration: combined heat/power/chilling solutions
- Power quality and reliability

Backup Generation

Some commercial and industrial customers stand to lose large amounts of money when their power is interrupted. This can be the effect of equipment shutting down suddenly, which can damage the equipment or from equipment stops ruining a whole batch of product. Also, some equipment takes a long time to restart once it is turned off by a power outage. Lost worker productivity can add to the expense for these customers.

Distributed generation is becoming a popular method of ensuring high power reliability for these types of customers.

Madison Gas & Electric Co. (MGE) in Wisconsin explored a variety of options for meeting peak capacity needs and for providing affordable backup service to customers with higher-than-average needs for service reliability before creating its own backup generation service (BGS) for customers with 75 kW or more demand by offering Cummins Onan PowerCommand generators. (Fig. 8-2)

MGE is an investor-owned public utility providing electric service to 123,000 customers in Dane County, WI. The utility began its BGS program in the spring of 1998. Previously, MGE offered interruptible rates and a financing program for customers to use to purchase their own generators. But MGE found customers didn't want to have the up-front costs of generator ownership nor the responsibility for maintenance and service.

Fig. 8-2 PowerCommand generators provide backup at Madison Gas & Electric.

These factors coupled with concerns about summer peak demand prompted the utility to start the BGS program.

The utility initially ordered 10 MW of capacity with hope of finding enough customers to use the generators in the summer. Within six months of the program's kickoff, MGE had signed a dozen customers, primarily through word-of-mouth advertising. Customers were contracted for three to 10 years at $1.5/kW per month for diesel and $3.50/kW per month for natural gas.

The distributed generation runs parallel with the utility. Each generator is located at the customer's business and has a radio link to the MGE dispatch office, allowing technicians at the utility to retrieve status information when a unit is running. In the event of a distribution outage, the generators switch on automatically with no more than 30 seconds of outage at the customer's business. Once grid power is restored, the generators synchronize with the utility's system and shut down without disrupting service.

"One MGE customer in the telecommunications industry has two 1,250 kW generators installed as backup because an outage can cost up to $1 million an hour," said Don Peterson, MGE senior director, Energy Products and Services. He said the utility has been surprised to see interest in the BGS plan from customers in service-related industries where there is normally not a higher-than-average need for reliability—including a landscaping business and a plumbing supply company.

PPL Corp. Distributes Fuel Cells

PPL Corp. became the first North American distributor of fuel cells manufactured by FuelCell Energy, Inc. during the fall of 2000. PPL also made an equity investment in FuelCell Energy of about $10 million, or roughly 1% of FuelCell's equity.

"This agreement is an important step forward for energy users because it significantly expands the range of energy solutions that we can provide," said William F. Hecht, chairman, president and chief executive officer of the Pennsylvania company.

A key part of PPL's energy solutions approach is distributed generation – providing small power generators to customers at their location. PPL offers a variety of distributed generation solutions through its GenSelect program, ranging from diesel generators to cutting-edge microturbines and fuel cells.

"Distributed generation presents an important opportunity for continued growth of PPL Corp.," said Hecht. "With the range of solutions we now offer, we clearly are a leader in the distributed generation business."

Fuel cells produce energy through a chemical reaction, rather than through combustion, by converting hydrogen and oxygen into electricity. This emerging technology has enormous potential to transform the traditional energy business.

Fuel cells are highly reliable and generate electricity in an environmentally responsible manner. Customers are expected to include hospitals, schools, data centers and other commercial and industrial electricity users.

"FuelCell Energy has a superior technology with a high probability of near-term commercialization," said Hecht. The company was selected after PPL's extensive study of the fuel cell industry. FuelCell Energy expects its units will be commercially available in 2001 and 2002.

Jerry Leitman, president and CEO of FuelCell Energy, said, "We're delighted to have PPL, a leading supplier of competitively priced electricity and energy services in the Mid-Atlantic region, as our first distribution partner in North America.

PPL expects to establish several demonstration sites for fuel cells over the next two years, Hecht said. The company already has distributor relationships with companies that manufacture other types of distributed generation.

PPL's position in distributed generation will be built upon its reputation as a marketer in competitive energy markets and brand recognition. Another significant factor is the company's network of profitable regional mechanical services companies, which will support the installation and service of fuel cells.

Case Studies

POWER FROM MINE GAS IN THE UK

These are testing times for Britain's domestic coal and gas industries. A recent announcement of a pit closure in Northumberland was yet another indication of the decline in coal mining in the UK. And power generators find it cheaper to import coal from Poland than to use domestic sources.

Gas suppliers have also been hit by the UK government's decision to ban new development in gas-fired power plants unless they show specific environmental benefits. And deregulation has not created the competitive market many had hoped for.

Against this background, with power producers reluctant to build new coal-fired plants and energy efficiency a hot topic, it takes innovative engineering to get a generating station off the drawing board. It is with some irony that a recent UK project uses coal mine resources to supply gas for power generation.

The gas supplied at this site is similar to coalbed methane, vented from closed deep coal mines. This gas was, in the past, vented to the atmosphere—a waste product from former coal mining operations. It is now being used as an energy resource.

Mine Gas

Two such projects were brought on stream in 1999 at abandoned collieries in the UK by specialist fuel supplier Coalgas. The first, at Markham in Derbyshire, supplies mine gas via a pipeline to a specialty chemical manufacturer. A second vent at Steetley, Nottinghamshire, fuels a power station owned by Independent Energy.

Both projects have significant environmental benefits, removing methane—a potent greenhouse gas—that would otherwise be vented to the atmosphere. Using the methane reduces the greenhouse impact of the gas by more than 95%.

The Steetley project was developed by Independent Energy, Coalgas, and turnkey contractor Wartsila NSD, and was funded as a commercial project by both Coalgas and Independent Energy. As a small-scale power station, Steetley is connected directly to the local distribution system, avoiding the transmission charges associated with using the national grid. With low fuel prices achieved through natural gas contracted at a price lower than that for pipeline natural gas and environmental benefits, Steetley is an attractive project.

The scheme is proving such a success that Coalgas is developing at least two other sites, and has begun investigations to quantify gas emissions for about 150 other identified vents and boreholes. It has received venture capital backing for these operations, and plans to have 40 mine gas extraction plants operating by the end of 2004. Coalgas reports that a number of large energy companies are expressing interest in licensing the technology ad exporting it to coal-rich markets such as China, India, or Eastern Europe.

Feasibility

It takes a lot of preparatory work, from feasibility studies and research, shaft surveys and equipment selection, to progress an identified site to one supplying fuel to a power station.

Coalgas' concept is to view abandoned mine shafts and old mine workings as large underground storage reservoirs. Monitoring of the Steetley and

Markham sites enabled information on barometric sensitivity, void space assessment and gas composition to be determined.

Coalgas carried out extensive research using mine plans, survey data, and geological reports. Long term monitoring was carried out using sensing and logging equipment. The results were integrated into a database and project economics weighed before a decision was made regarding project viability.

The consulting group IMC also carried out reserve evaluation studies for the mine gas project on behalf of Independent Energy, including an investigation into the extent of the mine workings, estimates of the gas that might have been produced from the coal seams during the life of the mine, and an estimate of reserves remaining and the rate at which they can be produced.

Coalgas has a license to extract the gas and has installed gas-pumping equipment and associated safety systems. Prior to this project the gas was vented naturally at a rate dependent on atmosphere pressure. To provide the gas at a constant rate for use in power generation it is necessary to use pumps to lower the pressure in the mine.

"Mine gas has a similar origin to coalbed methane but it contains more carbon dioxide and nitrogen as a result of coal seams being exposed to the atmosphere during long term operations," said David Oldham of Coalgas. "Methane concentrations in the gas vary from 60% to 80% and it is suitable as a fuel producing very low exhaust emissions. Its calorific value is approximately two-thirds of North Sea gas."

Planning permission for the Steetley site was obtained by Coalgas. The power plant is situated on a brownfield site near a major industrial complex. "Noise was a concern because of nearby houses but we has worked with Wartsila before on noise sensitive projects and knew we could overcome the problems by housing the engines in pre-cast concrete cells, and by paying particular attention to the design of the dump radiators," said Rob Jones of Independent Energy.

The fact that Steetley is one of the first projects of its kind in the UK was a hurdle to overcome. "There was therefore a need to educate and convince management and funders of the merits of the scheme," Jones said.

Prime Mover

Independent Energy could have considered gas turbines as the power plant's prime mover. However, there is no heat load and for simple generation, efficiencies exceeding 40% can be achieved with gas engines, compared with around 30% for similar-sized turbines. "The project would have been uneconomic at this level of efficiency and we did not even consider tendering gas turbines for this project," Jones said. "We also prefer gas engines because they are relatively straightforward to maintain."

Power at Steetley is generated by two 16V25SG Wartsila gas engines each rated at 3 MWe. Both units are able to run on mine gas but have the capability to run on natural gas. Wartsila was responsible for the plant design and takes care of the operation and maintenance for the site.

The Wartsila engine is a medium speed lean-burn gas engine for natural gas and coalbed methane. The engine bore is 250 mm and its stroke 300 mm. With 16 cylinders, the maximum electric output is 3.12 MWe and the electrical efficiency 40%.

It is built on a common base frame on which the engine and generator are assembled. For quick installation, some auxiliaries are mounted directly on the common frame such as the lube oil filter and lube oil coolers. The engine is equipped with the Wartsila NSD control system WECS 3000, which optimizes the engine performance with respect to power, efficiency and emissions.

In order to meet the methane content levels of around 70%, the gas engines have been modified. Due to the lower energy content, the fuel feed system has been rebuilt with

- increased gas pipe feeding system on the engine,
- new main gas valves,
- new solenoid driving electronic circuits,
- bigger gas regulating unit.

The gas pressure is increased in a compressor to about 4 bar. The volumetric flow necessary for the full 3 MW output has made it necessary to apply valves similar to Wartsila's bigger 285G and 34SG engines. To secure opening and closing of the solenoid driven main gas valves, the electronically driven coils require a more powerful current. The gas-regulating unit has also been increased to a larger size, similar to the Wartsila 285G engine.

"All these components have been tested in other projects without any problems occurring. The Steetley plant can operate on both natural gas and mine gas. The extraction rate at present is lower than anticipated, so gas is not produced rapidly enough to supply two engines – hence one runs on natural gas and the other on mine gas. Even when the methane content decreases, the engine can run full thanks to the modifications of the fuel system," said Thomas Stenhede of Wartsila.

Projected Performance

Although projected to run for many years, it is difficult to be precise about the reserves of mine gas because of the way in which the methane is produced. "Coal has a pore structure – rather like a sponge. The methane is held on the internal surfaces of the pores," said Oldham.

The methane has to travel through the interconnected pores and then into the fissures within the coal. It then migrates into the mine roadways before being pumped to the surface. "The methane therefore follows a very long and tortuous route and it is impossible analytically to establish how much is in connection with the surface and how rapidly it can be extracted."

The other complication with such a project is that the underground workings can flood with water seeping in from adjacent rock strata; as this water floods the underground roadways it may progressively block some of the gas flow paths.

A small-scale 3 MW power plant using mine gas as a base fuel has strong environmental benefits. The output of methane from a typical mine site in terms of CO_2 equivalent would be 307,000 tons a year. Using the gas as at Steetley would reduce this to about 4,500 tons a year CO_2 equivalent.

If this saving is projected to 50 abandoned gas sites, then by 2004 Coalgas estimates that its operations could save the equivalent of almost 400,000 tons of oil, which could translate into as much as 15% of renewable energy production in the UK.

Sunshine in Your Tank

There is an Amoco gas station in Maryland with a little something extra. Bobby Fletcher's station has an array of solar modules atop the pump island canopy, connected to the station's regular power supply. The array contributes directly to the station's bottom line by reducing the monthly electric bill.

The thin film solar modules generate up to 6 kW of electricity, enough to light the inside of the store and run some of its equipment.

The modules, manufactured by BP Solarex, are part of a BP Amoco project aimed at demonstrating real-world applications of solar electricity for homes and businesses.

"BP Amoco is committed to providing energy that helps reduce emissions while continuing to provide the fuels that ensure mobility for every," said Raymond Brasser, Amoco Atlantic senior vice president. "The solar modules at this station are evidence of how business needs for clean energy can be met while helping to reduce operating costs."

Amoco is working with the Maryland Energy Administration and BP Solarex to fund a permanent solar energy demonstration program at six elementary and middle schools. Each school will have a solar array and solar electric meter installed. To further promote solar energy education, special solar classroom curriculum is being made available to schools statewide. A special promotion was held in November, following the grand opening of the solar station, in which one penny (with a minimum of $2,000 guaranteed) for every gallon of gasoline pumped at Fletcher's gas station was donated to a solar education project at a nearby elementary school.

"Our success will ultimately be measured by how well we connect with our customers," said Brasser. "Projects like this that combine solar electric innovations at our stations, educational opportunities for our young people and support from our customers are a step in the right direction."

BP Amoco is considering the installation of solar panels at other service stations around the country.

THERMAL LOOP SYSTEM HEATS DOWNTOWN

Conectiv is building a thermal loop heating and cooling facility for the New Castle County Courthouse in Wilmington, Delaware. The Courthouse is the first customer to sign on to the system. Conectiv Thermal Systems is building the system to handle additional large customers in the area.

The development of the $15 million Wilmington District Thermal Energy Center can provide strong incentive for other businesses considering locating in Wilmington, according to Tom Shaw, Conectiv executive vice president for energy. Governor Thomas R. Carper expects the system to help revitalize the southern end of the city's central business district.

"The key to development of a thermal loop is the customers requiring new and upgraded facilities and our ability to provide thermal energy at a cost less than a customer can do it themselves," Shaw said.

District heating, with a central plant serving many nearby buildings, has been available for 100 years, but district cooling has only been available for the last 10 years.

"There is a good fit between this technology and the need for efficient and economical heating and cooling systems for mid-sized buildings," said Frank E. DiCola, Conectiv Thermal Systems vice president and general manager. "People are always favorable to ideas that will help them cut costs. It goes down to the economics." He estimates that customers who already have heating and cooling equipment can save 5 to 10 percent with district heating and cooling, but customers building new facilities can save up to 30 percent.

Construction is under way, with completion expected by April 2001. The existing heating and cooling equipment at Conectiv Corporate headquarters on King Street will be replaced with new hot and chilled water equipment. A 10-inch hot water line will provide 200 F water to heat the 12-story courthouse while a 24-inch line will carry 39 F water to the courthouse's air conditioners. The Conectiv building will also be heated and cooled by the new system.

"The courthouse will use about 30 to 35 percent of the capacity we can build at King Street," DiCola said. "If we ever sold out our capacity at King Street, we could build a second facility in the area and integrate the King Street facility with the new facility." The system was originally designed for expansion and integration in the future.

Water chillers will be housed in the basement of the Conectiv building, which had been a parking garage, and heaters will be sited in the penthouse.

The new cooling plant will have a 5,000-ton capacity. The hot water system will have three 16.8 MMBtu/hr heaters.

MICROTURBINE POWERS HOTEL

A La Quinta Inn and Suites located north of the Dallas/Ft. Worth airport is believed to be the first hotel in the country generating its own electricity with a microturbine. In 2000, La Quinta and TXU Energy Services began the test program to cut the hotel's peak demand by 63%. The hotel is a three-story, 140-room facility.

The self-contained 75 kW Honeywell microturbine is a quiet, natural gas-fired generating system. The microturbine automatically operates during periods when it can generate electricity for less than the local utility. The hotel has agreed to operate the turbine during voluntary curtailment.

"We estimate that by generating their own power, La Quinta will save up to $13,000 annually, depending on the number of hours they use the system," according to Randy Tipton of TXU Energy Services. "Distributed generation, such as this microturbine, has the potential to be of great benefit to energy users and producers. It improves power quality and relieves congestion during critical electricity distribution times."

Jim Ackles, La Quinta's Director of Energy and Engineering, says, "The ability to reduce our load by 50 to 70 percent within minutes is having a positive impact on our bottom line while helping the utility grid during peak use periods. Additionally, the Honeywell microturbine provides back up power in an emergency. And when we need it, it's simple and quick to start the microturbine. Either a TXU Energy Services employee or a La Quinta employee can get the microturbine going via modem."

Bill Brier, Edison Electric Institute vice president, notes, "A partnership such as this, between energy provider and energy user, is what electricity competition should be all about. Importantly, La Quinta lowers its energy bill. But perhaps more important, through the new energy options of TXU Energy Services, La Quinta also gains greater competitiveness in its own industry."

TXU Energy Services is a wholly owned subsidiary of TXU, a Dallas-based energy services company.

Edison Electric Institute (EEI) is the association of United States investor-owned electric utilities and industry affiliates and associates worldwide. TXU is an EEI member company.

Gas Turbines in the Oil Fields

PanCanadian Petroleum Limited is generating electricity at oil field locations by burning natural gas in four high-tech microturbines. The Capstone Turbine Corp. turbines each burn about 9,000 cubic feet of natural gas a day, generating up to 28 kW of power. Each turbine provides enough electricity to run two oil well screw pumps.

"This is a substantial step forward in our efforts to diversity the sources of our electricity, while at the same time conserving energy ad reducing emissions," said David Tuer, PanCanadian president and CEO.

PanCanadian and Capstone have worked together for several years to ensure that the microturbines were designed and built to meet the rigorous demands of Canada's oil fields. The turbines burn fuel that would otherwise be wasted. The electricity generated is used on site or sold into the power grid to increase the supply and reliability of Alberta's electrical system.

Capstone turbines have features that attracted PanCanadian to purchase the power plants. The turbines use air bearings, allowing them to spin at 96,000 revolutions per minute or 1,600 revolutions per second. At the same time, the air bearings substantially reduce the operating, maintenance and reconditioning expense of the generator. Conventional turbines are lubricated and cooled with oil, which requires more maintenance and increases maintenance costs. The Capstone turbines can burn a variety of fuels, including sweet natural gas, propane and sour gas with up to 7% hydrogen sulfide.

The PanCanadian installations achieve three principle environmental benefits. First, energy efficiency is increased by generating electricity from natural gas that would otherwise be flared at the well site. Second, with complete combustion of the gas, there are fewer hydrocarbons released that would occur in flaring. Third, the new on-site power reduces PanCanadian's demand for coal-fired power, which typically generates two times as many greenhouse gas emissions as natural gas-fired power generation.

EPRI buys EnAble Fuel Cell

EPRI's Power Electronics Applications Center (EPRI PEAC) Corp. purchased a 3 kW hydrogen-based power quality fuel cell system from EnAble Fuel Cells Corp. (FCC) in September of 2000. DCH Technology Inc., EnAble's parent company, issued a press release stating that EPRI PEAC had looked at a number of fuel cells on the market before choosing the EnAble system. According to the release, "They [EPRI] believe fuel cells – integrated with their own electric power management technologies – offer a clean, reliable, and responsive solution to power quality issues worldwide."

EnAble's Active Fuel Cell (AFC) systems range in size from 200 watts to 10 kW. That allows energy service providers to quickly adjust the system to handle varying load conditions.

The proton exchange membrane (PEM) fuel cell uses hydrogen, tap water, and air. The hydrogen enters the anode, where it separates into protons and electrons. The protons flow through the electrode, and the electrons travel through the external circuit to provide electrical power. At the cathode, the electrons and protons come together. They are then combined with oxygen from the air to produce water. Stephanie Hoffman, EnAble FCC's vice president and general manager, explained that the new system is an efficient and responsive solution to grid overload. "But that's just the tip of the potential of fuel cells in this market," she continued. "Fuel cell systems could help decentralize the power industry, protect the environment, and provide low-cost electricity for consumers."

Iowa Senior Living Facility Installs State's First Microturbine

Alliant Energy Resources, a subsidiary of Alliant Energy Corp., announced in the fall of 2000 that a residential and independent living facility in Toledo, Iowa, became the first test site in the state for small-scale power generation from a microturbine. Microturbines are a high-speed, quiet, low-emission technology that use natural gas, propane, or other fossil fuel to efficiently generate electricity.

Grandview Acres is using the microturbine to power the emergency panel for the facility. In addition, a heat recovery unit is capturing the excess heat produced by the microturbine and using it to provide hot water for use throughout the facility, as well as to preheat water for use in the boiler for space heating. As a result, Grandview Acres will save on their energy bill.

Jim Hoffman, president of Alliant Energy Resources said, "These types of distributed generation solutions are critical to the future of the energy industry. They offer clean and reliable power sources and commercial viable alternatives to upgrading transmission and distribution lines."

The microturbine Grandview Acres has selected is 28 kW, which is large enough to power about five average-sized homes. The cost to generate using the microturbine is approximately 6.5 cents per kilowatt-hour, fuel cost only. If combined heat and power (heat recovery) is used, the recovered heat credit lowers the gas price to between 3 and 4 cents per kilowatt-hour.

Terri Vaske, Administrator of Grandview Acres Toledo said, "We'll benefit through energy cost savings, but I think the most important benefit is going to be the insight that will be gained for future energy technology."

The Grandview Acres microturbine will be among the lowest NOx emissions of all combustion systems: less than nine parts per million. A traditional 30 kW diesel generator emits more nitrogen oxides in one hour than the microturbine unit does in more than nine days of full-load operation.

Bill Brier, EEI Vice President, Communication, said, "A partnership such as this, between energy provider and energy user, is a great example of the benefits of a changing industry. Importantly, Grandview Acres lowers its energy bill. But perhaps more important, through the new energy options of Alliant Energy Resources, Grandview Acres also gains greater competitiveness in its own industry."

Alliant Energy offers energy and environmental solutions in domestic and international markets through its regulated and diversified operations. Alliant Energy's utility subsidiaries serve more than 1.3 million electric, natural gas and water customers in Illinois, Iowa, Minnesota, and Wisconsin. Alliant Energy's international investments include power and combined heat and power facilities in China, New Zealand, Australia and Brazil. The company's diversified investment holdings include transportation, oil and gas development, real estate and telecommunications.

Wastewater Treatment Plant Installs Dozen Microturbines for Sewage Gas

The city of Allentown, PA, is adding a dozen, 30 kW microturbines to enable the city's wastewater treatment plant to generate its own electricity using the methane gas from the water treatment process as a fuel source. The city's energy services company, PPL EnergyPlus, a subsidiary of PPL Corp., will provide both fiscal and project management services. These services consist of the installation of a wide variety of energy conservation measures in addition to the microturbines.

Microturbines operate on a variety of sources of fuel, including natural gas, propane, bio-gas, and kerosene. In this case, the wastewater treatment plant will recycle the methane gas that is produced in the treatment process and use it as the fuel source for the microturbine. Both electricity and the waste heat produced by the microturbine will be recycled back into the treatment process. The facility produces approximately 80 million cubic feet of methane every year. This "free" fuel, combined with the energy efficient production of electricity inherent of a microturbine along with the recoverable heat and other energy conserving measures will produce significant energy saving for the city of Allentown. And, because microturbines produce ultra low emissions, less than 9 ppm of nitrogen oxides, the city of Allentown will also be helping the environment.

"This is an excellent opportunity for the City of Allentown to save money, improve our operations, and contribute to a better environment," said William Heydt, mayor of Allentown.

"Distributed generation — especially when it produces savings for the customer and benefits for the environment — can be a significant element of our portfolio of energy services for business," said Larry De Simone, president of PPL EnergyPlus. "We see clearly the potential of distributed generation and the benefits it can provide to our customers to meet their growing energy needs in a cost efficient manner."

Bill Brier, EEI Vice President, Communication, said, "A partnership such as this, between energy provider and energy user, is a great example of the benefits of a competitive electric industry. Importantly, the city of Allentown lowers its energy bill. But perhaps more important, through the new energy options of PPL Energy Plus, the benefits extend to all those who work and live in the city."

PPL EnergyPlus has worked with Allentown in the past on other energy-saving projects, including the installation of new lighting in city hall and improvements to other city-owned buildings. These projects are expected to save Allentown $300,000 over the next several years.

DukeSolutions and Harmony Products Developing Biomass-to-Energy Facilities

DukeSolutions and Harmony Products have agreed to jointly develop at least four additional animal-waste (biomass) processing plants, bringing the same waste-to-energy, organic fertilizer and environmental solutions to the

Southeast and Midwest that they are working to provide for the Virginia poultry industry and the surrounding Chesapeake watershed.

In late summer 2000, the companies announced plans to construct a $7 million Harrisonburg, VA, plant — America's first large-scale application of a poultry waste-to-energy-to-organic fertilizer operation.

Today's joint development agreement is a commitment to leverage at four-or-more additional facilities the technologies applied at the Harrisonburg plant. The sites planned for the Southeast and Midwest are undetermined. The companies also are exploring expanding the waste streams the plants could accept to include industrial wastewater sludge from wastewater processing plants in the poultry industry and other animal waste streams. Estimated construction costs for four new plants is $36 million. All are scheduled for completion by late 2001.

"We've seen estimates for up to 200 plants worldwide," said Keith G. Butler, DukeSolutions chief operating officer. "About half of which could be in the United States."

When completed, the plants will sell energy to industrial customers in the form of steam that is produced by the gasification of animal litter. DukeSolutions and Harmony Products will share ownership with Renewable Energy Corporation Limited of Sydney, Australia. DukeSolutions will focus on the waste-to-energy technology and operations and Harmony Products will operate the plants and market the fertilizer output.

Legislative activity on animal waste issues has been considered in 20 states. "Some people think a choice must be made between poultry jobs and a clean environment. We don't think so," said Butler. "Biomass-to-energy-to-fertilizer is a strategic alternative. Some see a ton of poultry litter and see negative environmental consequences and costs. We see a business plan and a tremendous opportunity to give something back to the world we live in. Each ton of litter processed equals about $100 in revenue in terms of fertilizer and energy. Each of these new plants could process up to 100,000 tons annually or enough energy to heat 15,000 homes."

The holistic approach the companies will utilize at the plants represents an entire business cycle. Harmony produces fertilizer used to grow feed for producers' animals whose wastes are used as raw material to help produce both fertilizer and fuel used in the production of fertilizer.

The DukeSolutions' developed biomass-to-energy plant would then allow Harmony to increase fertilizer output while providing an environmentally safe outlet for litter. "We can affix a price tag on these plants and a value to the profits we'll see," said Tom McCandlish, president of Harmony Products, "but how can you put a value on the Chesapeake Bay or contributing significantly to the worldwide sustainability?"

Under the joint development agreement, DukeSolutions and Harmony Products will cooperate in the analysis and development of a comprehensive solution for the beneficial use of animal waste streams. The intent is that optimum economic solutions should be adopted for management of a targeted organic waste stream. The solutions adopted should include both waste-to-energy components and fertilizer components in a proportion that is optimized based on local conditions including waste availability, fertilizer market, energy prices and other relevant factors.

"There are many companies trying to find a solution," said McCandlish. "We could have viewed Duke as a competitor and vice versa. But by optimizing both the energy and fertilizer we can offer a solution without added costs to our clients."

Animal waste has two potential values — fertilizer and energy. For instance, current industry practice for the disposal of poultry litter has been the bulk spreading of the waste materials over agricultural lands. Over-application of litter in geographically concentrated areas may deteriorate environmental quality. Poultry waste in bulk tends to decompose rapidly, releasing soil nutrients. When over applied, these water-soluble nutrients are dissolved in rainwater, which feeds into watersheds. Utilization of litter as fertilizer can lead to increased levels of phosphorus in the soil.

The demand for poultry is growing and so is the issue of poultry litter. It is becoming less valuable and harder to manage. There are numerous environmental and business benefits from biomass-to-energy-to-fertilizer projects. Decreased runoff of poultry litter into lakes and streams helps protect the environment. The gasification technology used to convert litter into fertilizer and energy produces the lowest emissions possible. The energy conserved means less reliance on fossil fuels. Additionally, a high-quality organic fertilizer is produced.

The organic fertilizer releases nutrients over time, reducing runoff when compared to traditional synthetic fertilizer. It also increases crop yields and quality, thus increasing land utilization and decreasing water consumption. The process of producing organic fertilizer is also environmentally friendly when compared to synthetics.

Harmony Products Inc., Chesapeake, VA, was organized by a group of fertilizer executives to develop, manufacture and market fertilizer and animal feed products. The company was started in February 1990, when the company was capitalized, and certain patent rights were acquired. The company has developed three generations of hydrolization and granulation technology that convert waste to fertilizer.

DukeSolutions helps customers make the most of their assets and increase cash flow and earnings by providing integrated energy solutions that reduce costs and improve energy efficiency. Serving industrial, commercial, institutional and governmental customers, DukeSolutions implements a full range of energy services including energy supply and logistics, reliability and risk management, on-site utility, information management and efficiency and productivity services. DukeSolutions is a wholly owned subsidiary of Duke Energy, a global energy company.

Elementary School Runs On 1,000 Watts of Solar

Wheelersburg Elementary School in Ohio added 1,000 Watts of solar power in the U.S. Department of Energy's (DOE) Million Solar Roofs Initiative. Ohio vaulted from last place in solar power in 1997 to number 14 in 2000. The southern Ohio school helped American Electric Power's (AEP) Learning from Light! program reach 90,000 Watts of installed capacity in Ohio as certified by DOE.

"Ohio does not come to mind when you consider solar energy," said John Hollback, AEP manager of environmental affairs in Ohio, who spoke at the dedication ceremony for the school's Solar Lab. "However, communities and schools across the state have embraced the Learning from Light! initiative, and momentum is building."

Glen Kizer, president of the Foundation for Environmental Education and AEP's major partner in Learning from Light!, concurs. "The program teaches local communities about the feasibility, reliability and economics of solar electricity and other ways to generate electricity," he said. "Everyone in Wheelersburg worked together to bring solar power to their school. More and more schools are finding out about our program and want to join."

In addition to AEP, sponsors of the school's solar project include the Ohio Department of Development's Office of Energy Efficiency, which provides funding to all such projects, and DOE's Million Solar Roofs Initiative. Students raised money from businesses and individuals to pay for the solar panel system and its connection to the school's electrical system.

"This system is an invaluable learning resource for the school and the community," Hollback said. "It's a springboard to understanding renewable energy and energy management and usage. Students, teachers and the public can go to AEP's Learning from Light! web site to see how much solar power is contributing to a school's electrical needs. They can determine how many solar panels, for example, it would take to keep the school running. This, in turn, helps them understand why it's important for a school's energy supplier, in this case AEP, to be able to use a variety of fuels to generate electricity."

"The solar school pages on the AEP web site also establish a network of teachers who can share information about ways the program helps improve students' knowledge. Schools are joining so fast that we haven't been able to add all of them to the web site yet," Kizer said.

AEP's Learning from Light! initiative promotes solar power projects anywhere in the world and offers advice to communities on how to take a system from concept to finished product. Local energy providers, government agencies, and environmental organizations can become partners. In the Wheelersburg project AEP engineers from Chillicothe helped the school select the best location for the solar panels, obtain them and install the system.

The Million Solar Roofs Initiative was announced in June 1997 to encourage the installation of solar energy systems at facilities of the U.S. government and on private buildings. The goal is to have one million solar roof systems installed by 2010.

American Electric Power is a multinational energy company based in Columbus, Ohio. AEP is one of the United States' largest generators of electricity with more than 38,000 MW of generating capacity. AEP is also one of the nation's leading wholesale energy marketers and traders. AEP delivers electricity to more than 4.8 million customers in 11 states - Arkansas, Indiana, Kentucky, Louisiana, Michigan, Ohio, Oklahoma, Tennessee, Texas, Virginia, and West Virginia. The company serves more than 4 million customers outside the United States through holdings in Australia, Brazil, China, Mexico and the United Kingdom. Wholly owned subsidiaries are involved in power engineering, construction, energy management, and telecommunications services.

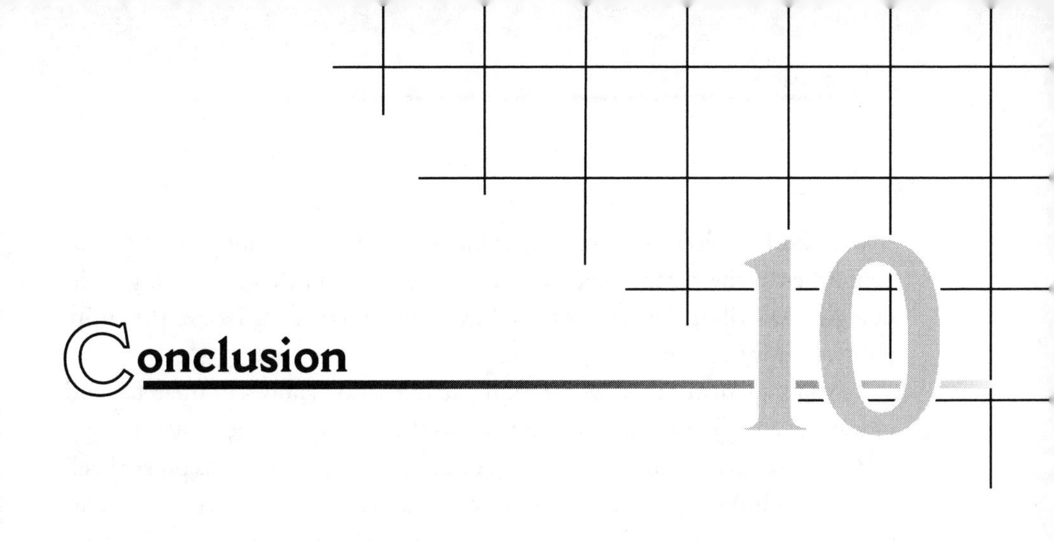

Conclusion

10

This is a very exciting time to be in the electric power industry, or in energy in general. Deregulation, technological innovation, and globalization are changing the rules of the game regularly. And one hot trend that is positioned to take advantage of today's tumultuous market is distributed generation.

With relatively low capital costs and short construction times, these technologies have much to offer. Small footprints and quiet workings lend themselves to easier permitting, especially in urban areas. Siting and permitting new generation is becoming ever more difficult, as communities unite to fight even relatively clean technologies such as combined-cycle natural gas-fired generation. People just don't want a big utility-scale generator sited anywhere near them, regardless of the tax dollars and good jobs that come with them. The public response of "not in my back yard," called NIMBY, is an increasing problem for generating companies. NIMBY can rear its ugly head to ruin projects seeking to create new, needed generation capacity or transmission and distribution capacity.

Distributed generation evades the NIMBY problem. People can build the technology they like to burn the fuel of their choice on their own land. Similarly, utilities can tuck some of these small generators into existing generation sites or into substations.

As we have seen, the term distributed generation includes a couple of established niche technologies and a handful of promising, emerging technologies. Distributed generation could very well change the face of the entire industry, if it lives up to its potential.

Installing distributed generation facilities can allow commercial and industrial energy users to take control of their energy usage and expenses, and even to recoup some of those expenses by selling their excess power back to the local utility or grid operator. It remains to be seen, however, when and if these technologies will make such a strategy economically attractive on a wide scale basis.

Distributed generation can also aid homeowners looking for more reliable electricity or electricity for rural locations such a farms or cabins. Distributed generation also holds promise for environmentally concerned consumers looking for a way to make a difference. Utilities are installing "green power" such as wind or geothermal and packaging it for these consumers, which can offset some of the dirtier power plants. Also, residential fuel cells look to be on the verge of commercial viability, and their green power would certainly allow homeowners to generate their own power in an environmentally harmless fashion.

Another hopeful arena for the distributed generation technologies, as we have seen here, is in grid support. Utilities can use small distributed generation facilities to supplement capacity at peak power demand times. This strategy aids reliability of today's overstressed grid and also helps offset the capital expenditure that would otherwise be required to build standard utility-scale facilities.

Since several of these technologies are compact enough to fit on large trucks or barges, distributed generation is also being used for emergency power when natural disasters strike. The applications for distributed generation are almost endless.

With industry research and development, government grants and tax incentives, and interest (and funding) from other industries, some of these technologies look like long-term winners.

But, as in all business, success in the end comes down to one major factor – economics. Currently, none of these technologies is economically attractive in a wide enough range of uses to merit the large commercial production needed to gain market share in the vast power industry. Microturbines are making inroads with their high efficiencies and low emissions.

Fuel cells also are making great strides toward more widespread use. The transportation industry is particularly interested in this technology, with most of the major automobile manufacturers sending large amounts of resources to one or another fuel cell manufacturing and research companies. There have been concept cars built to run on fuel cells, and several of our larger cities boast city buses running on fuel cell technology. Fuel cells have also benefited over the years from NASA's long-term interest.

In today's fluctuating market, it would be unwise to place a bet on the technology most likely to win the race to commercial success, regardless of the current odds. But one thing is certain – the race will be fascinating to watch.

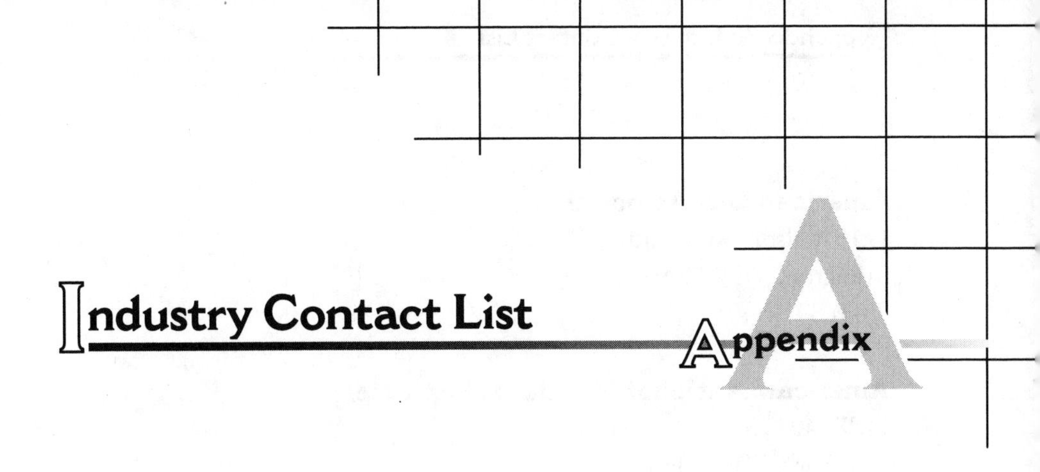

Industry Contact List

Appendix A

DOMESTIC

Allied Utility Network
1800 Peachtree St., NW, Ste. 401
Atlanta, GA 30367
(770) 972-0611

Alternative Fuels Data Center
P.O. Box 12316
Arlington, VA 22209
(800) 423-1363

American Boiler Manufacturers Association
950 N. Glebe Road, Suite 160
Arlington, VA 22203
(703) 522-7350

American Gas Association
1515 Wilson Boulevard
Arlington, VA 22209
(703) 841-8400

American National Standards Institute
11 W. 42nd St.
New York, NY 10036
(212) 642-4900

American Petroleum Institute
1220 L Street, NW
Washington, DC 20005
(202) 682-8000

American Society for Heating, Refrigerating and Air Conditioning
1791 Tullie Cr.
NE Atlanta, GA 30329
(404) 636-8400

American Society for Quality Control
Energy & Environmental Division
P.O. Box 3005
611 E. Wisconsin Ave.
Milwaukee, WI 53202
(414) 272-8575

American Solar Energy Society
2400 Central Avenue, Suite G-1
Boulder, CO 80301
(303) 443-3130

American Wind Energy Association
122 C Street, NW, 4th Floor
Washington, DC 20001
(202) 383-2500

ASME International
345 E. 47th Street
New York, NY 10017
(212) 705-7722

Association of Energy Engineers
4025 Pleasantdale Road, Suite 420
Atlanta, GA 30340
(770) 447-5083

Center for Renewable Energy And Sustainable Technology
1200 18th St. NW, Suite 900
Washington, DC 20036
(202) 530-2230

Cogeneration & Competitive Power Institute
c/o Association of Energy Engineers
4025 Pleasantdale Road, Suite 420
Atlanta, GA 30340
(770) 925-9633

Council on Environmental Quality
722 Jackson Place, NW
Washington, DC 20503
(202) 395-5750

Department of Energy, U.S.
1000 Independent Avenue, SW
Washington, DC 20585
(202) 622-2000

Distribution Systems Testing
Application and Research–DSTAR
GE-PSEC Bldg. 2, Rm. 501
One River Road
Schenectady, NY 12345
(518) 385-5370

Electric Generation Association
1401 H. Street, NW, Suite 760
Washington, DC 20005
(202) 789-7200

Electric Light & Power Magazine
1421 S. Sheridan Road
Tulsa, OK 74112
(918) 835-8161

Electrical Generating Association
2101 L Street, NW, Suite 405
Washington, DC 20037
(202) 965-1134

Electrical Generating Systems Association
10251 West Sample Road, Suite B
Coral Springs, FL 33065
(954) 755-AMPS

Electric Power Research Institute–EPRI
3412 Hillview Ave.
Palo Alto, CA 94304
(800) 313-3774

Electric Power Supply Association
1401 H. Street, NW, Suite 760
Washington, DC 20005
(202) 789-7200

Electrical Generating Systems Association–EGSA
1650 S. Dixie Highway, 5th Floor
Boca Raton, FL 33432
(561) 338-3495

Energy Federation Inc.
14 Tech Circle
Natick, MA 01760
(508) 653-4299

Energy Information Administration
Forrestal Building, Room 1F-048
Washington, DC 20585
(202) 586-8800

Energy Research Institute
6850 Rattlesnake Hammock Rd.
Naples, FL 34113
(941) 793-1260

Environmental Protection Agency
401 M St. SW
Washington, DC 20460
(202) 260-2090

Federal Energy Regulatory Commission
888 First Street, NE
Washington, DC 20426
(202) 208-0200

Geothermal Resources Council
2001 2nd Street, Box 1350
Davis, CA 95617
(916) 758-2360

IEEE Power Engineering Society
P.O. Box 1331
445 Hoes Lane
Piscataway, NJ 08855
(908) 981-0060

Institute of Electrical and Electronics Engineers Inc.
345 E. 47th Street
New York, NY 10017
(212) 705-7900

International Association for Energy Economics
28790 Chagrin Blvd., Suite 350
Cleveland, OH 44122
(216) 464-5365

International District Energy Association
1200 19th St. NW, Suite 300
Washington, DC 20036
(202) 429-5111

International Electrical Testing Association
PO Box 687
106 Stone Street
Morrison, CO 80465
(303) 697-8441

International League of Electrical Associations
2901 Metro Dive, Suite 203
Bloomington, MN 55425
(612) 854-4405

National Association of Electrical Distributors
1100 Corporate Square Dr., Suite 100
St. Louis, MO 63132
(314) 991-900

National Association of Energy Service Companies
1615 M Street, NW, #800
Washington, DC 20036
(202) 822-0950

National Association of
Power Engineers Inc.
One Springfield Street
Chicopee, MA 01013
(413) 592-6273

National Association of
Regulatory Utility Commissioners
P.O. Box 684
1100 Pennsylvania Ave, NW, Suite 603
Washington, DC 20044
(202) 898-2200

National Bioenergy
Industries Association
122 C Street, NW, 4th Floor
Washington, DC 20001
(202) 383-2540

National Electrical
Manufacturers Association
1300 N. 17th Street, Suite 1847
Rosslyn, VA 22209
(703) 841-3200

National Energy Information Center
1000 Independence Avenue, SW
Washington, DC 20585
(202) 586-8800

National Independent Energy Producers
601 13th Street, NW, Suite 320
Washington, DC 20005
(202) 793-6506

National Hydropower Association
122 C Street, NW, 4th Floor
Washington, DC 2001
(202) 383-2530

National Renewable Energy Laboratory
1617 Cole Boulevard
Golden, CO 80401
(303) 275-3000

National Science Foundation
4201 Wilson Boulevard
Arlington, VA22203
(703) 306-1224

North American Electric Reliability Council
Princeton Forrestal Village
116-390 Village Boulevard
Princeton, NJ 08540
(609) 452-8060

Occupational Safety and Health Administration
200 Constitution Avenue, NW, Room S2315
Washington, DC 20210
(202) 219-6091

Office of Energy Efficiency
And Renewable Energy
US Department of Energy
www.eren.doe.gov

Power Electronics
Applications Center
10521 Research Drive, Suite 400
Knoxville, TN 37932
(423) 974-8288

Power Engineering Magazine
1421 S. Sheridan Road
Tulsa, OK 74112
(918) 835-8161

Power Marketing Assoc.
1519 22nd Street, Suite 200
Arlington, VA 22202
(703) 892-0010

Public Utilities Risk
Management Association
352 Turnpike Rd., Suite 205
Southborough, MA 01772
(508) 624-6700

Solar Energy Industries
Association
122 C Street, NW, 4th Floor
Washington, DC 20001
(202) 383-2600

United States Energy Association
1620 Eye Street, NW, Suite 1000
Washington, DC 20006
(202) 331-0415

Utility Marketing Association
4755 Walnut St.
Boulder, CO 80301
(303) 786-7444

INTERNATIONAL

Australian Cogeneration Association
380 St. Kilda Rd.
P.O. Box 1469N
Melbourne, Victoria
Australia
+61 3/9696-8468

Canadian Electrical Association
1155 Metcalfe Street
Sun Life Building
Montreal, Quebec, Canada H3A 2V6
514/866-6121

Canadian Electricity Association
1 Westmount Square, Suite 1600
Montreal, Quebec, Canada H3Z 2P9
514/937-6181

Canadian Energy Research Institute
3512-33 St., NW, Suite 150
Calgary, Alberta, Canada T2L 2A6
403/282-1231

Canadian Standards Association
178 Rexdale Boulevard,
Etobicoke, Toronto, Canada M9W IR3
416/747-4000
800/473-6726

Central American Bank For Economic Integration
Apartado Postal 172
Tegucigalpa, DC, Honduras
504/228-2243

Cogen Europe
98 Rue Gulledelle
Brussels, Belgium
+32 2/772-8290

Electricity Association
30 Millbank
London, UK SWIP 4RD
+44 (0) 171/963-5700

Electricity Supply Association Of Australia Ltd.
Level 11
74 Castlereagh Street
Sydney, Australia NSW 2000
+61 2/9233-7222

European Bank for Reconstruction & Development
One Exchange Square
London, UK EC2A 2EH
44/0171-338-6499

European Investment Bank
100 Boulevard Konrad
Adenauer, 2950 Luxembourg
Luxembourg
352/4379-3146

Export Import Bank Of the U.S.
811 Vermont Ave., NW
Washington, DC 20571 USA

Finnish Energy Industries Federation, Finergy
P.O. Box 21
Etelaranta 10
Helsinki, Finland
+358 9/686-161

Industrial and Power Association
Brunel Building
Scottish Enterprise Technology Park
East Kilbride, Scotland, UK G75 0QD
+44 (0) 1355/272-630

Institution of Electrical Engineers
Michael Faraday House
Stevenage, Herts, UK SGI 2AY
+44 (0) 1438/767-249

International Energy Agency
9 Rue de la Federation
75739 Paris Cedex 15, France
331/4057-6551

International Finance Corp.
2121 Pennsylvania Ave., NW
Washington, DC, 20433 USA
202/473-1234

Inter-American Bank
1300 New York Ave., NW
Washington, DC, 20036 USA
202/623-1059

Latin American Association Of Development Financing Institutions
Paseo de la Republica
3211, Lima 27, Peru
511/442-2400

Norwegian Electric Power Research Institute
Sem Saelandsvei 11, N-7465
Trondheim, Norway
+47 73/59-7200

Secretariat for Central American Economic Integration
4a Avenida 10-25, Zona 14
Ciudad de Guatemala, Guatemala
+502 368/2151-54

South African Institute Of Electrical Engineering
P.O. Box 93541
Yeoville, 2143
Johannesburg, South Africa
+27 11/487-3003

Spanish Electricity Industry Association
Francisco Gervas No. 3
28020 Madrid, Spain
+1 941/793-1922

World Bank
3100 Massachusetts Ave. NW
Washington, DC 2008 USA
202/588-6692

World Energy Efficiency Association
1400 16th Street NW, Suite 340
Washington, DC 20036, USA
202/797-6570

World Energy Council
5th Floor, Regency House 1
4 Warwick Street
London, UK WIR 6LE
+44 20/7734-5996

World Trade Organization
Centre William Rappard
Rue de Lausanne 154
CH-1211 Geneva 21, Switzerland
+41 22/739-5111

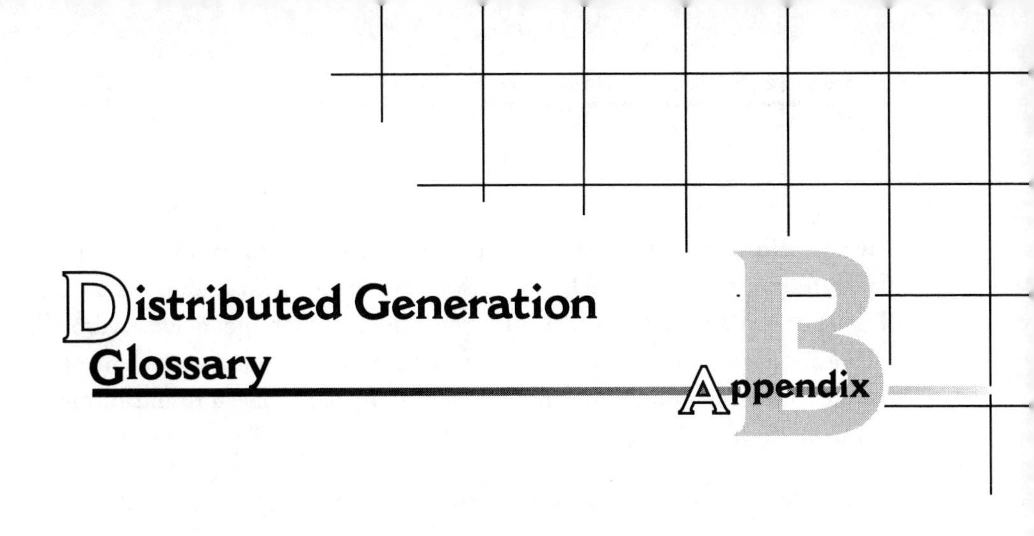

Distributed Generation Glossary

Appendix B

A

Affiliated power producer A company that generates power and is affiliated with a utility.

Aggregation The process of estimating demand and scheduling deliveries of power to a group of customers.

Aggregator A company that consolidates a number of individual users and/or supplies into a group. Marketing companies that pool gas from many sources into packages for resale to local distribution companies or end-users.

Air monitoring Intermittent or continuous testing of emission air for pollution levels.

Air pollution Contaminants in the atmosphere that have toxic characteristics and which are believed to be harmful to the health of animal or plant life.

Air pollution abatement equipment Equipment used to reduce or eliminate airborne pollutants, including particulate matter such as dust, smoke, fly ash, or dirt and sulfur oxides, nitrogen oxides, carbon monoxide, hydrocarbons, odors, and other pollutants. Examples of air pollution abatement structures and equipment include flue-gas particulate collectors and nitrogen-oxide control devices.

Air quality Air quality is determined by the amount of pollutants and contaminants present.

Air quality standards Pollutant limitations defined by law. Standards vary by country and are generally established by federal governments.

Alkaline fuel cell Fuel cells that use alkaline potassium hydroxide as the electrolyte. These are currently far too expensive for commercial use, but NASA uses them.

All-events contracts Also called "hell or high-water contracts." Sales or transport contracts requiring the customer to take or pay for contracted volumes or services, even if the seller is unable to deliver, regardless of who is at fault for the failure.

Allowable emissions The emissions rate of a stationary source calculated using the maximum rated capacity of the source and the most stringent applicable governmental standards.

Allowance An authorization to emit, during or after a specified calendar year, one ton of sulfur dioxide.

Allowance trading Buying or selling emission permits for sulfur dioxide. The Clean Air Act Amendments of 1990 require most fossil-fuel electric generating facilities to have allowances for each ton of sulfur dioxide emission produced. Utilities are allowed to buy and sell these allowances which the Environmental Protection Agency issues annually based on each facility's historic fuel usage and other factors.

Alternating current (ac) A periodic current, the average value of which over a period is zero. Unless distinctly specified otherwise, the term refers to a current that reverses its direction at regularly recurring intervals of time and that has alternately positive and negative values. Almost all electric utilities generate AC electricity because it can easily be transformed to higher or lower voltages.

Alternative fuel According to the stipulations of the Energy Information Administration, alternative fuels can include methanol; denatured ethanol and other alcohols; fuel mixtures containing 85% or more by volume of methanol, denatured ethanol, and other alcohols with gasolines or other fuels; liquefied petroleum gas (propane); hydrogen; coal derived liquid fuels; fuels other than alcohol derived from biological materials including biofuels such as soy-diesel; and electricity including electricity from solar energy. Also any other fuel which is substantially not petroleum and which would yield substantial energy security benefits and substantial environmental benefits. The term alternative fuel does not include alcohol or other blended portions of primarily petroleum-based fuels used as oxygenates or extenders or the 10% ethanol portion of gasohol.

Ambient conditions The outside weather conditions, including temperature, humidity, and barometric pressure. Ambient conditions can affect the available capacity of a power plant.

Amortize To spread out loan payments over a specified period of time.

Ampere A unit of measurement of electric current produced in a circuit by 1 Volt acting through a resistance of 1 Ohm.

Anchored merchant plant A power generating facility built with a partial power purchase agreement, guaranteeing some income for the project, but not sale of the entire capacity.

Ancillary services Services necessary to support the transmission of energy from resources to loads while maintaining reliable operation of the transmission provider's transmission system. Examples include voltage control and spinning dispatch.

Annual capacity factor The ratio between the actual heat input to a generating unit from an individual fuel or combination of fuels during a period of 12 consecutive calendar months and the potential heat input to the generating unit from all fuels if the unit had operated for 8,760 hours during that 12-month period at the maximum design capacity.

Annual demand The greatest electric power demand occurring in any calendar year.

Annual operating factor The annual fuel consumption divided by the product of design firing rate and hours of operation per year.

Annual operating time Also called annual service hours. The number of hours per year during which a unit or group of units is operated. Can be continuous or interrupted operations.

APPA American Public Power Association. The trade association of publicly held power entities.

Arms-length transactions Transactions between unaffiliated companies. For example, sales by a gas producer to an unrelated local distribution company.

Associated gas Natural gas found in association with crude oil, either as "dissolved" or "solution" gas within the oil-bearing strata or as "gas cap" gas just above the oil zone.

Attainment area A geographic area under the Clean Air Act which is in compliance with the Act's national Ambient Air Quality Standards. This designation is made on a pollution-specific basis.

Availability The unit of measure for the actual time a transmission line or generating unit is capable of providing service, if needed.

Available but not needed capability Net capability of main generating units that are operable but not considered necessary to carry load, and cannot be connected to load within 30 minutes.

Average revenue per kilowatt-hour The average revenue per kilowatt-hour of electricity sold by sector (residential, commercial, industrial, or other) and geographic area (State, Census division, and national), is calculated by dividing the total monthly revenue by the corresponding total monthly sales for each sector and geographic area.

Avoided cost A utility company's production or transmission cost avoided by conservation or purchasing from another source rather than by building a new generation facility.

B

Back-stopping Arranging for alternate supplies of gas in the event a user's primary source fails to be delivered.

Back-stops In financing lingo, guarantees or pledges of support that underpin contracts or financial agreements.

Back-up power Power supplied to a customer when its normal supply is interrupted.

Balance-sheet financing Using a corporate revenue stream as security to obtain debt financing for a project.

Barrel A volumetric unit of measure for crude oil and petroleum products equivalent to 42 U.S. gallons.

Baseload The minimum amount of electric power or natural gas delivered or required over a given period of time at a steady rate. The lowest load level during a utility's daily or annual cycle.

Baseload capacity The generating equipment normally operated to serve loads on an around-the-clock basis.

Baseload plant A plant, usually housing high-efficiency steam-electric units, which is normally operated to take all or part of the minimum load of a system, and which consequently produces electricity at an essentially constant rate and runs continuously. These units are operated to maximize system mechanical and thermal efficiency and minimize system operating costs.

Basis A geographic price differential between a particular market and the delivery point specified in an exchange-traded commodity contract.

Bcf Billion cubic feet.

Bid week A period in the latter part of each calendar month during which spot contracts are negotiated for gas to be delivered in the following calendar month.

Biomass Any body or accumulation of organic material. In the gas industry, biomass refers to the organic waste products of agricultural processing, feedlots, timber operations, or urban refuse from which methane can be derived.

Boiler A device for generating steam for power, processing, or heating purposes or for producing hot water for heating purposes or hot water supply. Heat from an external combustion source is transmitted to a fluid contained within the tubes in the boiler shell. This fluid is delivered to an end user at a desired pressure, temperature, and quality.

Bond A promise to pay a certain sum of money to the bond holder after a given period of time. Bond issuers borrow money from bond buyers for a pre-determined term, paying a pre-set interest rate.

Border price The price of gas at the U.S.-Canadian border, for the purpose of export/import licensing, taxation, or passthrough of costs in downstream sales.

Breaker A device to break the current of a given electric circuit by opening the circuit.

Btu British thermal unit. A standard unit for measuring the quantity of heat energy equal to the quantity of heat required to raise the temperature of pound of water by one degree Fahrenheit.

Bulk power The generation and high-voltage transmission of electricity.

Bundling 1) For electricity, combining the costs of generation, transmission, and distribution and other services into a single rate charged to the retail customer. 2) For natural gas, providing a combination of products and services in a single package at a fixed price with no customer ability to accept less than the entire package.

Burnertip The ultimate point where natural gas is used by the customer. The burnertip refers to any gas-fueled equipment, such as a furnace, cook top, or engine, used by the customer.

Bypass Direct sales by producers, pipelines, or marketers to end users, avoiding markups or transport fees of incumbent local distribution companies.

C

Capability The maximum load that a generating unit, generating station, or other electrical apparatus can carry under specified conditions for a given period of time without exceeding approved limits of temperature and stress.

Capacitor bank An assembly of capacitors and all necessary accessories, such as switching equipment, protective equipment, controls, and other devices needed for a complete operating installation.

Capacity The amount of electric power delivered or required for which a generator turbine, transformer, transmission circuit, station, or system is rated by the manufacturer.

Capacity charge An element in a two-part pricing method used in capacity transactions (energy charge is the other element). The capacity charge, sometimes called Demand Charge, is assessed on the amount of capacity being purchased.

Capital markets Venues for buying and selling investment instruments. Another sense of the term refers to the pool of money available from banks, institutional investors, and other sources of debt financing.

Carbon dioxide A colorless, odorless, nonpoisonous gas which occurs in ambient air. It is produced by fossil fuel combustion or the decay of materials.

Carbon monoxide A colorless, odorless, tasteless, but poisonous gas produced mainly from the incomplete combustion of fossil fuels.

Casinghead gas Natural gas that flows from an oil well along with the liquid petroleum. It is also called associated gas or solution gas because it resides beneath the earth's surface in conjunction with crude oil.

Chlorofluorocarbons A family of inert, nontoxic, easily liquefied chemicals used in refrigeration, air conditioning, packaging, and insulation, or as solvents or aerosol propellants. They are thought to be major contributors to potential ozone thinning and global warming and are therefore becoming increasingly regulated and their use more restricted.

Circuit A circuit is a conductor or system of conductors forming a closed path through which electric current flows.

City gate The physical interconnection of an interstate natural gas pipeline and the distribution system of a local gas utility. "Behind the city gate" refers to delivery points within the distribution system of the local gas utility.

Cogeneration The simultaneous production of power and thermal energy, such as burning natural gas to produce electricity and using the heat produced to create steam for industrial use.

Coincidental demand The sum of two or more demands that occur in the same time interval.

Coincidental peak load The sum of two or more peakloads that occur in the same time interval.

Combination pricing A pricing strategy in which aspects of cost, demand, and competition pricing methods are integrated.

Combined cycle An electric generating technology in which additional electricity is produced from otherwise lost waste heat exiting from the gas turbines. The exiting heat is routed to a conventional boiler or to a heat recovery steam generator for utilization by a steam turbine in the production of electricity. The process increases the efficiency of the electricity generating unit.

Combined-cycle unit An electric generating unit that consists of one or more combustion turbines and one or more boilers with a portion of the required energy input to the boiler(s) provided by the exhaust gas of the combustion turbine(s).

Combined utility A public utility, either privately owned or municipal, which sells both gas and electricity.

Commercial banks Financial institutions established for the purpose of lending money and conducting various financial transactions for third-party clients.

Commercial customers A statistical and regulatory category of energy use, embracing retail and wholesale trade, service establishments, hotels, offices, public institutions, and sometimes apartments that are separately metered.

Commercial entities As opposed to government entities or charities, commercial entities operate for the purpose of generating a profit.

Commercial financing Money obtained from banks and other commercial institutions.

Commercial operation Commercial operation begins when control of the loading of the generator is turned over to the system dispatcher. The amount of fuel used for gross generation, providing standby service, start-up and/or flame stabilization.

Commodities Good purchased frequently and in large quantities.

Commodity charge A customer charge for utility service that is proportional to the amount of gas or electricity actually purchased.

Common carrier A transporter obligated by law to provide service to all interested parties without discrimination to the limit of its capacity. If the capacity of a common-carrier pipeline is insufficient to satisfy demand, it must offer services "ratably" to all shippers in proportion to the amounts they tender for shipment.

Common purchaser An oil or gas carrier that is required by law to purchase without discrimination from all parties tendering oil or gas produced from a given reservoir, field, or area.

Competitive environment Rivalry among firms offering similar products or services to the same market.

Compressed natural gas Abbreviated CNG. Natural gas that is highly compressed, though not to the point of liquefaction, so that it can be used by an operation not attached to a fixed pipeline. CNG is used as a transport fuel.

Condensate Light hydrocarbon molecules that are liquid under atmospheric temperatures and pressures, and which are typically extracted from raw natural gas during processing.

Condenser Condensers are equipment in generating facilities that capture steam and turn it back into water for reuse in the feedwater system of the plant.

Consumption The amounts of fuel used for gross generation, providing standby service, start-up, and/or flame stabilization. May also be used to refer to customer use.

Contract carrier A transporter, such as a gas pipeline company, that provides its service on a discretionary, contractual basis for other parties.

Contract price Price of fuels marketed on a contract basis covering a period of one or more years. Contract prices reflect market conditions at the time the contract was negotiated and therefore remain constant throughout the life of the contract or are adjusted through escalation clauses. Generally, contract prices do not fluctuate widely.

Contract receipts Purchases based on a negotiated agreement that generally covers a period of one or more years.

Conventional gas Gas that can be produced under current technologies at a cost that is no higher than its current market value.

Convergence The coming together and merging of previously distinct industries. This phenomenon is currently under way for the electricity and fuels industries, particularly electricity and natural gas.

Cooling tower The portion of a power facility's water circulating system which extracts the heat from water coming out of the plant's condenser, cooling it down and transferring the heat into the air while the water returns through the system to become boiler make-up water.

Cooperative electric utility An electric utility legally established to be owned by and operated for the benefit of those using its service. The utility company will generate, transmit, and/or distribute supplies of electric energy to a specified area not being serviced by another utility. Such ventures are generally exempt from Federal income tax laws. Most electric cooperatives have been initially financed by the Rural Electrification Administration, U.S. Department of Agriculture.

Corporate bond A bond issued by a commercial entity.

Cost-of-service The paradigm of gas utility regulation in North America, whereby customer charges are based on the actual or forecast costs of providing the service, rather than allowing prices to rise to whatever customers may be willing to pay.

Counter-guarantee A guarantee of a guarantee. Multilateral agencies provide counter-guarantees to back-stop guarantees of performance issued by government agencies or government-owned utilities.

Country risks Risks involving the political, economic, or social climate of a given country.

Covenants Provisions of debt instruments that limit actions of the borrower that might increase risk of non-performance.

Credit rating A classification of the level of credit ascribed to a person, company, or country.

Credit support Agreements, guarantees, or financial instruments meant to improve the creditability of a venture.

Creditability The ability of a person, company, or country to obtain credit.

Credit worthiness Similar to creditability and credit rating, a company's credit-worthiness is a judgment of its trustworthiness and its ability to repay a loan.

Cryogenic Supercooled. Liquefied natural gas is cryogenically cooled for transport.

Cubic foot The most common unit of measurement of gas volume, it is the amount of gas required to fill a volume of one cubic foot under stated conditions of temperature, pressure, and water vapor.

Currency conversion Exchanging a country's currency for hard currencies traded in international monetary markets.

Current The flow of electrons in an electrical conductor. The rate of movement of the electricity, measured in amperes.

Customer density Number of customers in a given unit of area or on a given length of distribution line.

D

Debt service Payments made periodically to reduce a debt over time.

Declining block rate A utility rate structure by which customers who consume greater quantities are charged lower per-unit rates, which descend in a step-like fashion.

Dedication Legal reservation of reserves or production from a given property to a specific purpose or customer.

Deep gas Natural gas located 15,000 feet or more below the earth's surface.

Default Failure to make promised payments or fulfill promised obligations.

Deliverability The amount of gas that a pipeline or producer is capable of delivering, as limited by the terms of its supply contracts, its physical plant capacity, or by government regulations.

Demand Ability and willingness of customers to purchase a product or service. In electricity, the rate at which electric energy is delivered to or by a system, part of a system, or piece of equipment, at a given instant or averaged over any designated period of time.

Demand charge A customer charge for utility service that reflects the extent to which a particular customer chooses to purchase a right to draw a certain volume of gas at any time during the year.

Demand-side management (DSM) The term for all activities or programs undertaken by an electric system or its customers to influence the amount and timing of electricity use. Included in DSM are the planning, implementation, and monitoring of utility activities that are designed to influence consumers use of electricity in ways that will produce desired changes in a utility's load shape. These programs are dwindling, and expected to experience a great decline under deregulation.

Department of Energy Abbreviated DOE. Established in 1977, the DOE manages programs of research, development and commercialization for various energy technologies, and associated environmental, regulatory, and defense programs. DOE promulgates energy policies and acts as a principal adviser to the President on energy matters.

Deregulation Relaxing or eliminating laws and regulations controlling an industry or industries.

Direct current (dc) An electric current that flows in one direction with a magnitude that does not vary or that varies only slightly.

Derivatives Financial instruments whose values depend on those of other underlying assets. Examples include futures contracts, options, and swaps.

Direct purchases Purchases of gas by local distribution companies or end users, directly from producers rather than from merchant pipelines.

Disaggregation The breaking up of the traditional electric utility structure from a totally bundled service to an ala carte service.

Distribution automation A system consisting of line equipment, communications infrastructure, and information technology that is used to gather intelligence about the distribution system and provide analysis and control in order to maximize operating efficiency and reliability. It includes small distribution substations, sub-transmission and distribution feeder reclosers, regulators, and sectionalizers, which can be remotely monitored and controlled.

Distribution company An electric distribution company that provides only distribution services that are unbundled. Abbreviated DISCO.

Distribution system 1) For natural gas, the pipes and service equipment that carry or control the supply of natural gas from the point of local supply, or city gate, to the customer's meter. 2) For electricity, the substations, transformers, and lines that convey electricity from the generation site to the consumer.

E

Economic viability Having a reasonable chance of remaining financially solvent and generating a profit.

Edison Electric Institute The association of the investor-owned electric utilities in the United States and industry affiliates worldwide. Its U.S. members serve almost all of the customers served by the investor-owned segment of the electric utility industry. They generate almost 80% of all electricity generated by utilities and service more than 75% of all customers in the nation. EEI's basic objective is the "advancement in the public service of the art of producing, transmitting, and distributing electricity and the promotion of scientific research in such field." EEI compiles data and statistics relating to the industry and makes them available to member companies, the public, and government representatives.

Electric and magnetic fields Electric and magnetic fields are created when energy flows through an energized conductor. The electric field is from the voltage impressed on the conductors and the magnetic field is from the current in the conductors. These fields surround the conductors. Electric fields are measured in volts per meter or kilovolts per meter and magnetic fields are measured in gauss or tesla. Electric and magnetic fields occur naturally, but can also be created. There is debate regarding possible health effects of these fields when they occur in proximity to residences. Abbreviated EMF.

Electric capacity The ability of a power plant to produce a given output of electric energy at an instant in tie. Capacity is measured in kilowatts or megawatts.

Electric current A flow of electrons in an electrical conductor. The strength or rate of movement of the electricity is measured in amperes.

Electric rate schedule A statement of the electric rate and the terms and conditions governing its application, including attendant contract terms and conditions that have been accepted by a regulatory body with appropriate oversight authority.

Electric utility A corporation, person, agency, authority, or other legal entity or instrumentality that owns and/or operates facilities within the United States, its territories, or Puerto Rico for the generation, transmission, distribution, or sale of electric energy primarily for use by the public and files forms listed in the Code of Federal Regulations, Title 18, Part 141. Facilities that qualify as cogenerators or small power producers under the Public Utility Regulatory Policies Act (PURPA) are not considered electric utilities.

Electricity The flow of electrons in a conducting material. The flow is called a current.

Emissions Any waste products leaving a power plant. This term generally applies to air pollution, but it can also apply to soil or water waste issues. There are many substances that can be emitted from power plants, and most of them are regulated and monitored.

End user The ultimate consumer, as opposed to a customer purchasing for resale.

Energy charge The portion of the charge for electric services that is based on the electric energy either consumed or billed.

Energy deliveries Energy generated by one electric utility system and delivered to another system through one or more transmission lines.

Energy efficiency Refers to programs that are aimed at reducing the energy used by specific end use devices and systems, typically without affecting the services provided. These programs reduce overall electricity consumption (reported in megawatt-hours), often without explicit consideration for the timing of program-induced savings. Such savings are generally achieved by substituting technically more advanced equipment to produce the same level of end-use services (e.g., lighting, heating, motor drive) with less electricity. Examples include high-efficiency appliances, efficient lighting programs, high-efficiency heating, ventilating and air conditioning (HVAC) systems or control modifications, efficient building design, advanced electric motor drives, and heat recovery systems.

Energy marketer An entity, regulated by the Federal Energy Regulatory Commission, which arranges bulk power transactions for end users. The main goal of energy marketers is determining the best overall fuel choice for customers, whether it be natural gas, electricity, oil, etc., and then delivering that fuel to the customer. They deal in the open market, taking full title to the energy until they resell it to an end user.

Energy Policy Act of 1992 Legislation which authorized FERC to introduce competition at the wholesale level through new open access requirements for transmission and authorizing exempt wholesale generators.

Energy receipts Energy generated by one electric utility system and received by another system through one or more transmission lines.

Energy Power is the capability of doing work. Energy is power supplied over time, expressed in kilowatt-hours. Energy can take on different forms, some of which are easily convertible and can be changed to another form useful for work. Most of the world's convertible energy comes from fossil fuels that are burned to produce heat that is then used as a transfer medium to medium to mechanical or other means in order to accomplish tasks. Electrical energy is usually measured in kilowatt-hours, while heat energy is generally measured in British thermal units.

Enhanced recovery A family of technologies applied to extract more of the hydrocarbons in place in a reservoir than can be produced by natural fluid pressure.

Environmental Protection Agency EPA. This agency administers federal environmental policies, enforces environmental laws and regulations, performs research, and provides information on environmental subjects.

Equity Ownership right in a property. Equity sponsors own a share of a project, as opposed to debt sponsors, which lend money to the project.

Exempt wholesale generator A company that generates power solely for wholesale use and does not sell it to the public. They are exempt from PUHCA.

Expatriation of funds The ability to take funds out of a country.

Export credit agencies Organizations established to support the export of goods and services from a given country by providing loans and other types of financial support.

Exposure Risk of losing money.

Expropriation Governmental seizure of assets against the owner's will.

Externalities Factors affecting human welfare not included in the monetary cost of a product, such as air pollution caused by power generation.

F

Federal electric utilities A classification of utilities that applies to those that are agencies of the federal government involved in the generation and/or transmission of electricity. Most of the electricity generated by federal electric utilities is sold at wholesale prices to local government-owned and cooperatively owned utilities, and to investor-owned utilities. These government agencies are the Army Corps of Engineers and the Bureau of Reclamation, which generate electricity at federally owned hydroelectric projects. The Tennessee Valley Authority produces and transmits electricity in the Tennessee Valley region.

Federal Energy Regulatory Commission (FERC) A quasi-independent regulatory agency within the Department of Energy having jurisdiction over interstate electricity sales, wholesale electric rates, hydroelectric licensing, natural gas pricing, oil pipeline rates, and gas pipeline certification.

Federal Power Act Enacted in 1920, and amended in 1935, the Act consists of three parts. The first part incorporated the Federal Water Power Act administered by the former Federal Power Commission, whose activities were confined almost entirely to licensing non-Federal hydroelectric projects. Parts II and III were added with the passage of the Public Utility Act. These parts extended the Act's jurisdiction to include regulating the interstate transmission of electrical energy and rates for its sale as wholesale in interstate commerce. The Federal Energy Regulatory Commission is now charged with the administration of this law.

Federal Power Commission The predecessor agency of the Federal Energy Regulatory Commission. The Federal Power Commission (FPC) was created by an Act of Congress under the Federal Water Power Act on June 10, 1920. It was charged originally with regulating the electric power and natural gas industries. The FPC was abolished on September 20, 1977, when the Department of Energy was created. The functions of the FPC were divided between the Department of Energy and the Federal Energy Regulatory Commission.

Feedwater The water used in the boiler system of generating plants. It is treated to make it as pure as economically feasible to keep the boiler clean and operating properly.

Financial closing The time at which all necessary financial commitments have been made in the legal sense.

Firm gas Gas sold on a continuous and generally long-term contract.

Firm power Power or power-producing capacity intended to be available at all times during the period covered by a guaranteed commitment to deliver, even under adverse conditions.

Firm service Sales and/or transportation service provided without interruption throughout the year. Firm services are generally provided under filed rate tariffs.

Fixed cost An expense that does not change in response to varied production levels. Also called imbedded cost.

Force majeure A contractual provision by which a party's obligations are waived if a superior force, such as weather, war or an act of God, makes it impossible for those obligations to be met.

Foreign direct investment Capital investments made in a country by foreign companies.

Forward contract A contract for future delivery at a price determined in advance.

Franchise A special privilege conferred by a government on an individual or a corporation to engage in a specified line of business, or to use public ways and streets.

Fuel cell A device capable of converting natural gas, hydrogen, or other gaseous fuels directly into electricity and heat via an electrochemical process that avoids the energy losses associated with combustion and the spinning or reciprocation of mechanical parts.

Fuel expenses Costs include the fuel used in the production of steam and/or electricity at an electric power plant. Other associated costs include unloading the shipped fuel and all handling of the fuel up to the point where it enters the power plant. Fuel expenses are generally the largest expense category for electric power generating facilities.

Futures A derivative financial instrument that creates a contractual obligation to buy or sell a specified volume of an underlying commodity at a set price on some future date.

G

Gas marketer/broker A non-regulated, competitive buyer/seller of natural gas. The marketer/broker may be an aggregator.

Gas processing Processing raw gas to remove liquid hydrocarbons such as propane and butane, toxic or corrosive substances such as hydrogen sulfide and carbon dioxide, and adjust the residue gas to a standard heating value.

Gas turbine Consists of an axial-flow air compressor and one or more combustion chambers where liquid or gaseous fuel is burned The hot gases that are produced are passed to the turbine and where the gases expand to drive the generator and are then used to run the compressor.

Gas A fuel burned under boilers and by internal combustion engines for electric generation. These include natural, manufactured, and waste gas.

Generating unit Any combination of generators, reactors, boilers, combustion turbines, or other prime movers operated together or physically connected to produce electric power.

Generation The process of producing electric energy by transforming other forms of energy. It also refers to the amount of electric energy produced, generally expressed in kilowatt-hours or megawatt-hours.

Generator A machine that converts mechanical energy into electrical energy.

Generator nameplate capacity The full-load continuous rating of a generator, prime mover, or other electric power production equipment under specific conditions as designated by the manufacturer. Installed generator nameplate rating is usually indicated on a plate physically attached to the generator.

Geothermal plant An electric power plant in which the prime mover is a steam turbine. The turbine is driven either by steam produced from hot water or by natural steam that derives its energy from heat found in rocks or fluids at various depths beneath the surface of the earth. The energy is extracted by drilling and/or pumping.

Gigawatt Abbreviated GW. A unit of electric power equal to one billion watts or one thousand megawatts.

Gigawatt-hour Abbreviated GWh. One billion watt-hours.

Global warming A hypothesized increase in worldwide atmospheric temperatures, caused by an intensification of the natural greenhouse effect thought to attend the socially accelerated accumulation of carbon dioxide and other heat-retaining gases.

Green marketing Using an ecological perspective in marketing, packaging, or promoting a product as environmentally benign or beneficial.

Greenfield plant Refers to a new electric power generating facility built from the ground up on a site that has not been used for industrial uses previously. Essentially a plant that starts with a green field. Plants that are built on sites that have already been used for another power plant or other industrial use are called brownfield plants.

Greenhouse effect The natural warming of the earth's lower atmosphere, associated with solar energy reflected from the surface and retained by water vapor and irradiative gases, including carbon dioxide and methane.

Greenhouse gases Those gases, such as carbon dioxide, nitrous oxide, and methane, that are transparent to solar radiation but opaque to longwave radiation. Their action in the atmosphere is similar to that of glass in a greenhouse.

Grid The layout of an electrical distribution system.

Gross generation The total amount of electric energy produced by the generating units at a generating station or stations, measured at the generator terminals.

Ground A conducting connection, whether intentional or accidental, by which an electric circuit or equipment is connected to the earth, to some conducting body of relatively large extent that serves in place of the earth.

H

Hard currency Currencies that are readily traded in open exchanges. As opposed to soft currencies, the value of a hard currency is set by market mechanisms and not by official decree.

Heat rate A power plant term for the efficiency of the power plant. Heat rate measures how much of the fuel that is burned actually turns into electricity. Heat rate is generally represented as a mixture of British and metric units, Btu/kWh.

Heating value The amount of heat produced by the complete combustion of a unit quantity of fuel. Heating value for gas is specified as either "gross" or "net" depending primarily upon whether adjustment is made for the latent heat of vaporization of the combustion products.

Hedging Strategies to protect against financial loss resulting from an unfavorable price change, by locking in or containing the price of a future transaction. Hedging strategies for commodities include the purchase and sale of futures contracts and other derivatives.

Hertz The international standard unit of frequency, defined as the frequency of a periodic phenomenon with a period of one second. Abbreviated Hz. Electricity is generally either 50 Hz or 60 Hz.

Hubs A set of nearby interconnections between two or more pipelines and/or local distribution company main lines, and sometimes storage facilities, configured and operated to facilitate arms-length sales and purchases.

Hybrid structures Financing structures that employ features of two or more financing approaches, such as non-recourse debt, export credit financing, and public debt market refinancing.

Hydrocarbon An organic chemical compound of hydrogen and carbon in either gaseous, liquid, or solid phase. The molecular structure of hydrocarbon compounds varies from the simple, such as methane, to the very heavy and very complex.

I

Independent power producer (IPP) A company that generates power but is not affiliated with an electric utility.

Industrial market Companies that buy products or services for business or trade use.

Industrial sector Electric utilities generally divide customers into classes, broadly, residential, commercial and industrial. The industrial sector includes manufacturing, construction, mining, agriculture and others. Industrial users generally have heavier electrical use than residential or commercial users.

Institutional investor A company that takes interest, either debt or equity, in ventures sponsored by third parties.

Intermediate load In an electric system, intermediate load refers to the range from base load to a point between base load and peak load. This particular stage may be the mid-point, a percent of the peak load, or the load over a specified time period.

Internal combustion plant A plant in which the prime mover is an internal combustion engine. This type of engine has one or more cylinders, in which the process of combustion takes place, converting energy released from the rapid burning of a fuel-air mixture into mechanical energy. Diesel or gasoline engines are the principal types used in electric plants. These plants are generally used during periods of high electricity demand as peaking facilities.

Interruptible gas Gas sold to customers with a provision that permits curtailment or cessation of service at the discretion of the distributing company under certain circumstances, as specified in the service contract.

Interruptible load Refers to program activities that, in accordance with contractual arrangements, can interrupt consumer load at times of seasonal peak load by direct control of the utility system operator or by action of the consumer at the direct request of the system operator. It usually involves commercial and industrial consumers.

Interruptible service Sales and transportation service that is offered at both a lower cost and lower level of reliability. Under this service, gas companies can interrupt customers on short notice, typically during peak service days in the winter. Interruptible services are provided through individually negotiated contracts and, in most cases, the price and availability charged take into account the price of the customer's alternative fuel.

Investment bank An organization that assists prospective project sponsors in locating and securing debt and equity financing.

Investor-owned utility (IOU) Electric utilities organized as tax-paying businesses and generally financed by the sale of securities. The properties are managed by shareholder-elected representatives. These are usually set up as publicly owned corporations.

J

Joule A measurement of energy. It is the work done by a force of one Newton, when the point at which the force is applied is displace one meter in the direction of the force. It is equal to 0.239 calories. In electrical theory, one joule equals one watt-second.

K

Kilovolt Equal to 1,000 volts. Abbreviated kV.

Kilowatt Abbreviated kW. A measurement of electric power equal to one thousand watts. Electric power capacity of 1 kW is sufficient to light 10, 100-watt lightbulbs.

Kilowatt-hour (kWh) A measure for energy that s equal to the amount of work dome by 1,000 watts for one hour. Consumers are charged for electricity in cents per kilowatt hour. One kilowatt hour is enough electricity to run 10, 1,000-Watt light bulbs for one hour. Abbreviated kWh.

L

Liquid assets Valuable property that can be readily traded for cash in open markets.

Liquidated damages A calculation of penalties and expenses incurred when a contractor fails to fulfill the specifications of the contract. The contractor is liable to pay liquidated damages to the project sponsor in the event of failure to meet the contract specifications.

Liquidity The efficiency, including ease, speed, and economy, with which a good can be bought or sold.

Liquefied natural gas Methane that is chilled below its boiling point so it can be stored in liquid form, thereby occupying 1/625 of the space it requires at ambient temperatures and pressures.

Load The amount of electric power required at a given time by energy consumers which can be divided into three major classes — industrial load, commercial load, and residential load.

Local distribution company A utility that owns and operates a natural gas distribution system for the delivery of gas supplies from interstate pipelines at the city gate to the customer. Abbreviated LDC.

M

Manufactured gas Energy-rich vapors produced from controlled thermal decomposition or distillation of hydrocarbon feedstocks, including coal, oil, and coke-oven feedstocks. Manufactured gas historically had a low heating content. In the late 1970s and 1980s, however, the U.S. Department of Energy promoted construction of facilities to manufacture high-Btu synthetic natural gas suitable for commingling with natural gas in transmission pipelines.

Market-clearing price The price at which supply and demand are in balance with respect to a particular commodity at a particular time. A market-clearing price is high enough to prevent a shortage but low enough to ensure that all supplies then available can be sold.

Marginal cost The cost to increase output by one unit, such as the cost to produce one additional kWh of electricity.

Marginal revenue Change in total revenue from the sale of one additional product or unit.

Maximum demand The greatest of all demands of the load that has occurred within a specified period of time.

Mcf One thousand cubic feet. One Mcf of natural gas has a heating value of approximately one million Btu, also written MMBtu.

Megawatt Abbreviated MW. One million watts.

Megawatt-hour Abbreviated MWh. One million watts for one hour.

Merchant plant An electricity generating facility built and operated without long-term contracts guaranteeing sale of the electricity generated. Many such facilities are partial merchant plants, with contracts guaranteeing sale of a certain percentage of generation to a nearby utility.

Methane The simplest, lightest gaseous hydrocarbon, it is the primary component of natural gas.

Microturbine A small gas turbine used for electric power generation.

Molten carbonate fuel cells These fuel cells run a relatively high temperatures and are very fuel flexible.

Monopoly The exclusive control of a commodity or service by one entity. In the gas industry, interstate pipelines and local distribution companies are generally monopolies. Electricity has traditionally been operated as a local monopoly. Even after deregulation, it is anticipated that transmission infrastructure will remain regulated monopolies.

Multilateral agencies Organizations funded by more than two countries, established to support economic and social development. Multilateral agencies such as the World Bank, the Asian Development Bank, the Inter-American Development Bank, and the European Bank for Reconstruction and Development, provide direct loans and credit support for loans from other institutions, including performance guarantees.

Municipal utility An electric utility system owned and/or operated by a municipality that generates and/or purchases electricity at wholesale for distribution to retail customers generally within the boundaries of the municipality.

N

Natural gas A naturally occurring mixture of hydrocarbon and nonhydrocarbon gases found in porous geological formations beneath the earth's surface, often in association with petroleum. The principal constituent is methane.

Natural gas liquids Abbreviated NGL. Hydrocarbon components of wet gas whose molecules are larger than methane but smaller than crude oil. Gas liquids include ethane, propane, and butane.

Net capability The maximum load-carrying ability of the equipment, exclusive of station use, under specified conditions for a given time interval, independent of the characteristics of the load. Capability is determined by design characteristics, physical conditions, prime mover, energy supply, and operating limitations, such as cooling and circulating water supply and temperature, headwater and tailwater elevations, and electrical use.

Net generation Gross generation less the electric energy consumed at the generating stations for station use.

Net-present-value The lifetime worth of an asset, calculated at the present time.

Net summer capability The steady hourly output, which generating equipment is expected to supply to system load exclusive of auxiliary power, as demonstrated by tests at the time of summer peak demand.

Net winter capability The steady hourly output which generating equipment is expected to supply to system load exclusive of auxiliary power, as demonstrated by tests at the time of winter peak demand.

Nonattainment area A geographic region in the United States designated by the Environmental Protection Agency as having ambient air concentrations of one or more criteria pollutants that exceed National Ambient Air Quality Standards.

Noncoincidental peak load The sum of two or more peakloads on individual systems that do not occur in the same time interval. Meaningful only when considering loads within a limited period of time, such as a day, week, month, a heating or cooling season, and usually for not more than one year.

Non-firm power Power or power-producing capacity supplied or available under a commitment having limited or no assured availability.

Non-recourse financing Debt financing structures with security for loan repayments assigned to project revenues and assets rather than sponsor's corporate balance sheets.

Nonutility generator Abbreviated NUG. A facility that produces electric power and sells it to an electric utility, usually under long-term contract. NUGs also tend to sell thermal energy and electricity to a nearby industrial customer.

Nonutility power producer A corporation, person, agency, authority, or other legal entity that owns electric generating capacity and is not an electric utility. Nonutility power producers include qualifying small power producers and cogenerators without a designated franchised service territory.

North American Electric Reliability Council Abbreviated NERC. Electric utilities formed NERC to coordinate, promote, and communicate about the reliability of their generation and transmission systems. NERC reviews the overall reliability of existing and planned generation systems, sets reliability standards, and gathers data on demand, availability, and performance.

O

Off-peak gas Gas that is to be delivered and taken on demand when demand is not at its peak.

Off-peak power Power supplied during designate periods of relatively low system demands.

Ohm The unit of measurement of electrical resistance. Specifically, an Ohm is the resistance of a circuit in which a potential difference of one volt produces a current of one Ampere.

Open access Access to the commodity market via unbundled transmission capacity, for producers, end users, local distribution companies, and other gas resellers, on substantially equal terms for all kinds of shippers.

Operator The legal entity, usually a working interest owner, responsible for the management and day-to-day operation of a well or lease.

Options Financial derivatives that convey a right, but not an obligation, to buy or sell an underlying asset at a specified price until some fixed deadline, at which time the right expires.

Outage The period during which a generating unit, transmission line, or other facility if out of service.

Over-the-counter Pertaining to a financial asset or commodity, bought and sold away from an organized exchange, and thus an asset over which parties negotiate and conduct transactions directly between themselves.

Ozone A compound consisting of three oxygen atoms. It is the primary constituent of smog.

Ozone transport Ozone transport occurs when emissions from one area drift downwind and mix with local emissions contributing to the ozone concentrations in the downwind area.

P

Peak days In electricity, the days in the summer months when the demand for electricity is at its highest level due to air conditioning load. For natural gas, peak days are the days in the winter months when demand for gas is at its highest level due to most heating equipment being used.

Peak load plant A plant usually housing old, low-efficiency steam units; gas turbines; diesels; or pumped-storage hydroelectric equipment normally used during the peak-load periods.

Peaking capacity Capacity of generating equipment normally reserved for operation during the hours of highest daily, weekly, or seasonal loads. Some generating equipment may be operated at certain times as peaking capacity and at other times to serve loads on an around-the-clock basis.

Performance guarantee A promise to accept responsibility for another party's obligations. Multilateral agencies often issue performance guarantees of a state-owned utility's obligation to pay for power generated by an IPP.

Phosphoric acid fuel cells The only commercially available type of fuel cell, these offer very high efficiencies in cogeneration uses. They have a slow start-up time so they are not suited to emergency power needs.

Photovoltaic cell A type of semiconductor device in which the absorption of light energy creates a separation of electrical charges.

Pipeline All physical equipment through which gas is moved in transportation, including pipes, valves, and other attachments.

Pipeline quality gas Natural gas within 5% of the heating value of pure methane, or 1,010 Btu per cubic foot under standard atmospheric conditions, and free of water and toxic or corrosive contaminants.

Plant-use electricity The electric energy used in the operation of a plant. This energy total is subtracted from the gross energy production of the plant; for reporting purposes the plant energy production is then reported as a net figure. The energy required for pumping at pumped-storage plants is, by definition, subtracted, and the energy production for these plants is then reported as a net figure.

Polymer electrolyte membrane fuel cells Also called proton exchange membrane fuel cells. These are typically developed for smaller applications, such as light-duty vehicles, small buildings, and for electronics such as video cameras and laptop computers.

Power marketer A company that buys and resells power. These merchants typically do not own generating facilities.

Power The instantaneous current being delivered at a given voltage, measured in watts, or more usually kilowatts. Power delivered for a period of time is energy, measured in kilowatt-hours.

Prime mover The engine, turbine, water wheel, or similar machine that drives an electrical generator. Generally, a prime mover refers to a device that converts energy to electricity directly, such as photovoltaic solar and fuel cells.

Private power producer Any entity that engages in wholesale power generation or in self-generation.

Privatization The sale or transfer to private individuals or businesses of assets or businesses owned by the government or the conversion of a government-owned firm or industry to private ownership.

Pro-forma A general budget describing the basic economic expectations of a venture.

Producing capacity The maximum rate at which a field or some other producing unit can flow hydrocarbons through existing surface equipment without causing damage to the productive reservoir.

Producing sector The part of the gas industry that finds hydrocarbons, conveys them from the reservoir to the surface, and delivers them to a buyer in a first sale.

Project finance Debt financing structures with security for loan repayments assigned to project revenues and assets rather than sponsor's corporate balance sheets.

Public debt Loan funds that originate from institutions other than commercial lenders or other traditional finance institutions. Examples include revenue bonds purchased by individual investors or corporations.

Public equity Equity shareholdings that are sold and traded in public stock markets.

Public utility Publicly owned electric utilities are nonprofit local government agencies established to serve their communities and nearby consumers at cost, returning excess funds to the consumer in the form of community contributions, economic and efficient facilities, and lower rates. Publicly owned electric utilities number approximately 2,000 in the United States, and include municipals, public power districts, state authorities, irrigation districts, and others.

PUC Public utility commission. An administrative or quasi-judicial body at the state provincial or municipal level, whose functions include regulation of public utilities.

PUHCA The Public Utility Holding Company Act of 1935. PUHCA regulated the large interstate holding companies that monopolized the electric utility industry in the early part of the twentieth century. Recently, the Securities and Exchange Commission has been interpreting this legislation more leniently, allowing foreign firms to buy domestic utilities and allowing non-utility companies to purchase utilities without becoming registered holding companies. It is broadly believed that as deregulation of the electric industry becomes more comprehensive, PURPA will either be repealed or replaced.

Purchased power adjustment A clause in a rate schedule that provides for adjustments to the bill when energy from another electric system is acquired and it varies from a specified unit base amount.

PURPA The Public Utility Regulatory Policy Act of 1978. PURPA promotes energy efficiency and increased use of alternative energy sources, encouraging companies to build cogeneration facilities and renewable energy projects. Facilities meeting PURPA requirements are called qualifying facilities or QFs.

Q

Quad Abbreviation for one quadrillion Btu. For natural gas, this is roughly one trillion cubic feet.

Qualifying facility A generator that 1) qualifies as a cogenerator or small power producer under PURPA and 2) has obtained certification from FERC. They generally sell power to utilities at the utilities' avoided cost. Abbreviated QF.

R

Rate base The value of property upon which a utility is permitted to earn a specified rate of return as established by a regulatory authority. The rate base generally represents the value of property used by the utility in providing service and may be calculated by any one or a combination of the following accounting methods: fair value, prudent investment, reproduction cost, or original cost. Depending on which method is used, the rate base includes cash, working capital, materials and supplies, and deductions for accumulated provisions for depreciation, contributions in aid of construction, customer advances for construction, accumulated deferred income taxes, and accumulated deferred investment tax credits.

Raw gas Natural gas as it issues from the reservoir, including mainly methane and possibly heavier hydrocarbons, as well as impurities such as hydrogen sulfide, carbon dioxide, and water.

Recovery rate The fraction of the original oil or gas in place deemed to be recoverable with current technology. Alternatively the fraction of original oil or gas in place projected to be recovered with installed or firmly planned field equipment.

Regional transmission organization A voluntary organization of transmission owners, transmission users, and other entities approved by the Federal Energy Regulatory Commission to efficiently coordinate transmission planning and expansion, operation, and use on a regional basis.

Regulation The government function of controlling or directing economic entities through the process of rulemaking and adjudication.

Regulatory environment Regulations and enforcement affecting marketing activities that are laid down by government and non-government entities.

Renewable energy Refers to any source of energy that is constantly replenished through natural processes. Sunlight, moving water, geothermal springs, biomass, and wind are all examples of renewable energy resources used to generate electricity.

Reserve margin (Operating) The amount of unused available capability of an electric power system at peakload for a utility system as a percentage of total capability.

Reserves With respect to oil or gas, that part of the resource that is commercially recoverable under current economic conditions with current technology. "Proved" reserves are the portion of the resource that is in known reservoirs and believed to be recoverable with the highest degree of confidence. "Indicated" or "probable" reserves are the additional resources associated with known reservoirs that are expected to be recoverable. "Speculative" reserves are those resources, in addition to the others already mentioned, outside the vicinity of known reservoirs, which are expected to be recoverable.

Residential The residential sector is defined as private household establishments which consume energy primarily for space heating, water heating, air conditioning, lighting, refrigeration, cooking, and clothes drying. The classification of an individual consumer's account, where the use is both residential and commercial, is based on principal use. For the residential class, do not duplicate consumer accounts due to multiple metering for special services (water, heating, etc.). Apartment houses are also included.

Restructuring The process of separating, or unbundling, the true monopoly functions of a local natural gas utility — such as the physical delivery, or distribution, of natural gas to a home or business through pipelines — from those services — such as providing natural gas supply — that can be offered competitively.

Retail wheeling The transmission of power to an individual customer from a generator of electricity other than the host utility. The National Energy Policy Act, enacted in 1992, prohibits the Federal Energy Regulatory Commission from mandating retail wheeling. States and their regulatory bodies, however, are free to enact their own retail wheeling initiatives.

Retail Sales covering electrical energy supplied for residential, commercial, and industrial end-use purposes. Other small classes, such as agriculture and street lighting, also are included in this category.

Retrofit To change an existing piece of equipment or facility in order to improve its performance or efficiency.

Rule 144a A rule of the U.S. Securities and Exchange Commission providing for exemption from certain reporting laws for qualifying debt issues. Institutional investors, not private individuals, can purchase bonds issued under Rule 144a.

Running and quick-start capability The net capability of generating units that carry load or have quick-start capability. In general, quick-start capability refers to generating units that can be available for load within a 30-minute period.

Rural electric cooperatives (REC) Organizations composed of rural customers that band together to generate or purchase power at wholesale rates and then distribute it at retail rates. Also called co-ops.

S

Sales for resale Energy supplied to other electric utilities, cooperatives, municipalities, and Federal and State electric agencies for resale to ultimate consumers.

Secondary markets Resale markets for goods whose first-sale prices and/or allocation are constrained by long-term contracts, monopoly power, or government regulation. Unconstrained secondary-market transactions can reallocate goods to their highest value uses, mitigate or even eliminate shortages and surpluses caused by price controls, and thus substantially improve the efficiency of resource allocation.

Self-regulation Industry activities and efforts to police itself.

Service territory The geographical area served by a particular utility company.

Small power producer Under the Public Utility Regulatory Policies Act (PURPA) a small power production facility or small power producer generates electricity using waste, renewable, or geothermal energy as a primary energy source. Fossil fuels can be used, but renewable resources must provide at least 75% of the total energy input.

Solid oxide fuel cell These can be differentiated from other fuel cell types by their high operating temperature and solid state ceramic cell structure. These may someday be used in combined-cycle fuel cell hybrid applications with high efficiencies.

Spinning reserve That reserve generating capacity running at a zero load and synchronized to the electric system.

Spot market transaction Commodity sale/purchase transactions whereby participants' buy and sell commitments are of short duration at a single volumetric price, relative to term or contract markets in which transactions are long-term and pricing provisions are often complex. For natural gas, spot transactions typically have durations of one month or less.

Standby facility A facility that supports a utility system and is generally running under no-load. It is available to replace or supplement a facility normally in service.

Standby service Support service that is available, as needed, to supplement a consumer, a utility system, or to another utility if a schedule or an agreement authorizes the transaction. The service is not regularly used.

Steam electric plant A plant in which the prime mover is a steam turbine. The steam used to drive the turbine is produced in a boiler where fossil fuels are burned.

Stranded costs This refers to a utility's fixed costs, usually related to investments in generation facilities, that would no longer be paid by customers through their rates in the event that they opt to purchase power from other suppliers.

Swaps Over-the-counter financial derivatives in which a buyer and seller of a physical commodity or financial asset exchange cash flows from physical transactions. They make or receive periodic payments to or from one another based on the difference between physical market realizations and a specified index price.

T

Take-or-pay A contractual obligation to pay for a certain threshold quantity of gas whether or not the buyer finds it possible to take timely full delivery. Typically, the buyer still retains a right to take the volumes for which it have prepaid, but only after taking all the volumes it had a subsequent obligation to buy.

Tariff A list of terms, conditions, and rate information applied to various types of gas service. These tariffs are filed and approved by the Federal Energy Regulatory Commission or the state regulator.

Technology risks Risks associated with equipment technology. For example, a new type of gas turbine that has not been fully proven in service might be considered a technology risk, as opposed to a turbine that has operated successfully in the field for years.

Tenor Time period under which a loan is to be repaid.

Therm One therm equals 100,000 Btu.

Time-of-day pricing A rate structure that prices electricity at different rates, reflecting the changes in the utility's costs of providing electricity at different times of the day.

Tolling An arrangement whereby a party moves fuel to a power generator and received kilowatt hours in return for a pre-established fee.

Tranche When financing for a project has several different parts with different terms, each part is called a tranche. For example, commercial bank loans with 7-year terms are one tranche, while export credits with 12-year terms and sponsors equity are two other tranches.

Transmission The movement or transfer of electric energy over an interconnected group of lines and associated equipment between points of supply and points at which it is transformed for delivery to consumers, or is delivered to other electric systems. Transmission is considered to end when the energy is transformed for distribution to the consumer. Also, in natural gas, the conveyance of natural gas from producing to consuming areas through large-diameter, high-pressure pipelines.

Transmission grid The high voltage wires that connect generation facilities with distribution facilities. It is the infrastructure through which power moves around the United States. It is necessary to carefully coordinate use of the transmission system to ensure reliable and efficient service.

Transmission line A set of conductors, insulators, supporting structures, and associated equipment used to move large quantities of power at high voltage.

Transmission system An interconnected group of electric transmission lines and associated equipment for moving or transferring electric energy in bulk between points of supply and points at which it is transformed for delivery over the distribution lines to consumers or is delivered to other electric systems.

Transportation Service in which a gas pipeline or distribution company moves gas owned by others from one location to another for a fee.

Turbine A machine for generating rotary mechanical power from the energy of a stream of fluid (such as water, steam, or hot gas). Turbines convert the kinetic energy of fluids to mechanical energy through the principles of impulse and reaction, or a mixture of the two.

Turnkey contract A contract under which a company agrees to engineer, procure equipment for, construct and start up a facility, and assumes liability for meeting targeted dates and prices. When it is finished, the contractor hands the operating facility over to the owner.

U

Ultra-high voltage systems Electric systems in which the operating voltage levels have a maximum root-mean-square ac voltage above 800,000 volts (800 kV).

Unbundling The process of separating natural gas services into components with each component priced separately. Traditionally, numerous gas services, such as sales, local transportation, and storage, had been tied together and offered to customers as a single, bundled product. By separating services into components, unbundling enables customers to compare the value of each service to its price. Unbundling also allows customers to choose those individual services that meet their own energy needs. This practice is expected to be part of the deregulation of the electric industry as well with generation, transmission, and distribution segments separated, as well as various value-added services and ancillary services.

Usage rates Segmenting consumers according to the volume of product they buy and the speed at which they use it.

Utility Privately owned companies and public agencies engaged in the generation, transmission, or distribution of electric power for public use.

V

Value-added services Services, such as security monitoring, telecommunications, internet access, and others, that add value to electric services. Other services which can be offered by utilities to achieve greater customer satisfaction and loyalty.

Variable costs Those costs borne by electric utilities that vary with the level of electric output and include fuel expenses.

Vertical disaggregation Separating electric generation, transmission, and distribution functions of a utility into separate companies.

Vertically integrated utility Utilities that sell power on a bundled basis and whose activities run the full range of different functional activities of generation, transmission, and distribution. With deregulation of the electric utility industry well under way, vertically integrated electric utilities may well be on their way out.

Volt The measure of pressure that pushes electric current through a circuit.

Volt-ampere reactive A reactive load, typically inductive from electric motors, which cases more current to flow in the distribution network than is actually consumed by the load. This required excess capability on the generation side and causes greater power losses in the distribution network. Abbreviated VAR.

W

Waste-to-energy plants A steam-turbine generating facility that uses municipal solid waste as the primary energy source to produce the steam used in the generating process.

Watt The basic expression of electrical power or the rate of electrical work. One watt is the power resulting from the dissipation of one joule of energy in one second.

Watt-hour An electrical energy unit of measure equal to one watt of power supplied to, or taken from, an electric circuit steadily for one hour.

Wellhead price The price or value in the first sale of oil or gas, for regulatory, royalty, or tax purposes. This first sale may actually take place at the well, the lease or unit boundary, the tailgate of a gas-processing plant, or the intake flange of a pipeline.

Wheeling service The movement of electricity from one system to another over transmission facilities of intervening systems. Wheeling service contracts can be established between two or more systems.

Wheeling The transportation of power to customers. Wholesale wheeling is transmitting bulk power over the grid to power companies. Retail wheeling is transmitting power to end users, such as homes, businesses, and factories.

Wholesale wheeling The use of transmission facilities of one system to transmit power by agreement of and for another system with a corresponding wheeling charge Wholesale wheeling involves only sales for resale and occurs when the buyer of the power resells the wheeled power to retail customers.

Index

A

B

C

D

E

F

G

H

I

J

K

L

M

N

O

P

Q

R

S

T

U

V

W

Also by Ann Chambers

Merchant Power: A Basic Guide

Natural Gas and Electric Power in Nontechnical Language

Power Primer: A Nontechnical Guide from generation to End-Use

Power Branding

Co-author of

Power Industry Desk Reference CD

Power Industry Dictionary

Power Industry Abbreviator